SCC Library

P9-CKH-322

Santiago Canyon College
Library

BUILDING A RESILIENT TOMORROW

QC
903.2
.U6
H55
2020

BUILDING A RESILIENT TOMORROW

How to Prepare for the Coming Climate Disruption

Alice C. Hill

AND

Leonardo Martinez-Diaz

OXFORD
UNIVERSITY PRESS

Santiago Canyon College
Library

on1084573626

2/10/20 #31 Amazon LW

OXFORD
UNIVERSITY PRESS

Oxford University Press is a department of the University of Oxford. It furthers
the University's objective of excellence in research, scholarship, and education
by publishing worldwide. Oxford is a registered trade mark of Oxford University
Press in the UK and certain other countries.

Published in the United States of America by Oxford University Press
198 Madison Avenue, New York, NY 10016, United States of America.

© Oxford University Press 2020

All rights reserved. No part of this publication may be reproduced, stored in
a retrieval system, or transmitted, in any form or by any means, without the
prior permission in writing of Oxford University Press, or as expressly permitted
by law, by license, or under terms agreed with the appropriate reproduction
rights organization. Inquiries concerning reproduction outside the scope of the
above should be sent to the Rights Department, Oxford University Press, at the
address above.

You must not circulate this work in any other form
and you must impose this same condition on any acquirer.

CIP data is on file at the Library of Congress
ISBN 978–0–19–090934–5

3 5 7 9 8 6 4 2

Printed by Sheridan Books, Inc., United States of America

Alice dedicates this book to Peter, Liza, and Julia
Leonardo dedicates this book to Aravind, Cristina,
and José Antonio

CONTENTS

CONTENTS

PART III
THE UPENDERS

ACKNOWLEDGMENTS

This book benefited directly from the generous guidance and advice of numerous people, to whom we are deeply indebted. Deep thanks go to Kate Daly and Danny Morris, who provided strategic edits to strengthen the entire text. We owe a debt of gratitude to those who generously agreed to read all or parts of the manuscript and provided valuable feedback: Samuel Adams, Jeff Alexander, John Balbus, Brandon Bedford, Frank Femia, Eric Letvin, Cristina Martínez, Megan McCaslin, A. R. Siders, Peter Starr, Liza Starr, Julia Starr, and Caitlin Werrell. In addition, we are grateful to Gabriel Vincent Kahn, Bill Kakenmaster, and Stephanie Strazisar for research assistance, careful editing, and moral support throughout the process. We also want to thank our editor at Oxford University Press, Sarah Humphreville, who was an enthusiastic guide and kept us headed in the right direction.

Many experts graciously gave us their time to broaden our understanding of resilience efforts on the ground, including David Adams, Steven Bingler, Daniel Kreeger, Christine Morris, Ann Phillips, and Cheryl Rosenbaum. In writing this book, we benefited from insights we have absorbed over the years from a community of friends and colleagues devoted to building resilience to climate change. This community incudes Amal-Lee Amin, Terry Anderson, Isaac Anthony, Vicki Arroyo, Adrienne Arsht, Bilal Ayyub, Jack Baker, Jainey Bavishi, John Bluedorn, Catherine Berg, Sharon Burke, Ana Campos García, Christina Chan, Jane Chang, Daniel Clarke, Rachel Cleetus, Ryan Colker, John Conger, Heather Conley, Matthew Cranford, Roger-Mark De Souza, Craig Davies, Geoffrey Dabelko, Brian Deese, Rowan Douglas, Tamara Dickinson, Stanislas Dupré, Sophie Evans, Aaron Ezroj, Shiloh Fetzek, Chris Field, John Firth, Christopher Flavelle, George Frampton, Linda Fried, Craig Fugate, Gerry Galloway, Michael Gerrard, Francis Ghesquierre, Sherri Goodman, Jessica Grannis, Henry Green, Katharine Jacobs, Dave Jones, Sarah Jordaan, Elizabeth Hadly, Stéphane Hallegatte, Carl Hedde, Mark Heising, Claude Henry, John Holdren, Niels Holm-Nielsen, Saleemul Huq, Jesse Michael Keenan, Jay Koh, Fred Krupp, Howard Kunreuther, Anthony Kuczynski, Michael Kuperberg, Robert Lempert, Gary Libecap, Meghan Linkin, Amy Luers, Darío Luna Plá, Katherine Mach, Olivier Mahul, Samir Malviya, Lenny Marcus, Douglas Mason, John Kerry, Eric McNulty, Erwann Michel-Kerjan, Richard Moss, Senator Lisa Murkowski, Frank Nutter, Milo Pearson, Laura Petes, John Podesta, Curtis Ravenel, John Roome, Lea Rosenblum, Susan Ruffo, George Schultz, Richard Sorkin, Roger Sorkin, Rod Schoonover, Daniel Stander, Daniel Stadtmüller, Skip Stiles, Chittayong Jao Surakitbanharn, Stacy Swann, James Sweeney, Dave Titley, Frances Ulmer, Shalini Vajjhala, Oscar Vela Treviño, Nikhil da Victoria Lobo, Ahmad Wani, Emily Wasley, Bill Whalen, Kathleen White, Senator Sheldon Whitehouse, Roy Wright, Michelle Wyman, Simon Young, Craig Zamuda, Daniel Zarrilli, Paul Zukunft, and Astrid Zwick.

We also want to express our gratitude to our mentors in the Obama Administration who gave us extraordinary opportunities to serve the public. For Alice, they are Lisa Monaco, Janet Napolitano, and Susan Rice. For Leonardo, they are Lael Brainard, Marisa Lago, and Susan Reichle.

Alice thanks Thomas Gilligan and the Hoover Institution, as well as the Rockefeller Foundation's Bellagio Center, for providing the support and time to think deeply about resilience and to collaborate with other scholars. Leonardo thanks the community of experts at the World Resources Institute for providing a rich environment for discussion and reflection on climate resilience. The views expressed in this book are the authors' own and do not necessarily reflect the views of the Hoover Institution or the World Resources Institute.

Finally, we want to thank our families and our partners for their emotional support and intellectual partnership during the writing of the book. Without their love and generosity, we would have never completed this project.

INTRODUCTION

Building a Resilient Tomorrow

For most of her career, retired US Navy Rear Admiral Ann Phillips never thought she would one day work full time on climate change. "I mean, I watched Al Gore's movie for the first time two years ago," she recalled in 2018. (Gore's movie, *An Inconvenient Truth*, had come out a dozen years before.) Phillips had largely spent her thirty-one-year career in active service on naval bases, from Guam to San Diego to Portugal. For three decades, her focus was on traditional Navy priorities such as driving ships and participating in joint naval exercises with allied forces.

But now here she was, poring over maps of projected storm surge in Norfolk, Virginia. Pulled into an ad hoc Navy task force, she and her colleagues had been charged with identifying how climate change impacts might affect naval installations and operations. What the maps revealed was not good. A city built at the intersection of rivers and the ocean nearly three centuries ago, Norfolk is used to living with water. Like many cities all over the world, Norfolk has used its historical experience with water to plan for flooding. But with climate change, the past will no longer be a reliable guide to the future. "When we looked at the storm surge mapping for Virginia, and we saw what was covered with water, we realized we had a problem," Phillips remembers.

What Phillips and her colleagues understood is that because of sinking land and rising seas, Norfolk faces the fastest rate of sea-level rise along the Atlantic coastline of the United States. Over the past century, the city has experienced over 18 inches (46 cm) of sea-level rise—about twice the global average. Experts attribute about eight inches (20 cm) of that rise to climate change,[1] and they expect Norfolk's water line to rise an additional 24 inches (61 cm) by mid-century.[2] Compounding the problem is the growing threat of intensifying Atlantic storms.

Norfolk's relentless sea-level rise keeps some US military and political leaders awake at night, because this is no ordinary coastal town. "The most important place in the country you've never heard of," Phillips calls it. Norfolk and the surrounding area house some twenty-nine military facilities, including the largest naval base in the world and a strategic command headquarters for the North Atlantic Treaty Organization (NATO). Tens of thousands of active-duty personnel live there. America's aircraft carriers are built in Norfolk, as are some of its nuclear submarines. Many US Navy and Army ships are harbored there. Surrounding the seafaring complex is a dense network of railroads, roads, and shipyards connecting residential, commercial, and government buildings. As the rising waters threaten to flood docks, damage facilities, make roads impassable, and drive residents from their homes, this key military hub is at risk.

Like Ann Phillips, Christine Morris was no climate change expert. She studied international affairs at George Washington University and worked for over a decade in the nonprofit sector, focusing mostly on housing and education policy. When she was working as an assistant to Norfolk's city manager in 2013, her boss asked her to apply for the job of Norfolk's chief resilience officer, whose main responsibility is to prepare the city for climate change impacts. Morris demurred. "I'm not sure I'm the right person for this," she told her boss. "I'm not an engineer or an architect." But Morris could do what the position demanded, which was to bring together experts from many different sectors and find solutions, so she got the job.

Morris's responsibility was to worry about a different part of Norfolk. Despite its military might, the city suffers from severe poverty. A fifth of Norfolk's 250,000 residents live below the nation's poverty line; the percentage of residents in poverty is twice Virginia's average. A history of racial discrimination still haunts the city. Poverty became such a pressing concern that in 2013, the mayor convened a special commission to find ways to address it.

Not far from Norfolk's vast military installations is Tidewater Gardens, a low-income public-housing complex that was built in the 1950s. The median household income of residents is $12,000 a year. "The folks there are outside the cone of opportunity," Morris puts it delicately. As with other parts of Norfolk, Tidewater Gardens is at severe risk of flooding. The entire complex will be torn down and redeveloped with flood-protection features. If this is done right, Morris thinks it could be an opportunity not just for these residents of public housing but for the entire city. Norfolk could build resilience against climate impacts and address the lack of economic opportunity at the same time. "We must use the flooding as an opportunity to frame the city we want to be," she says.

This book is about people like Ann Phillips and Christine Morris, professionals who, in the course of pursuing careers unrelated to climate change, found themselves pulled into key roles in helping their communities to build resilience against this great challenge of our time. The book is also about places like Norfolk, where climate change impacts will upend lives and livelihoods unless communities build resilience first. Indeed, Norfolk will appear frequently throughout the book because its story captures, in a concentrated dose, the struggles and dilemmas that many communities will face when building resilience to climate change. The city's experience illustrates this volume's key themes, including the implications of climate change for national security, financial markets, public finance, migration, economic inequality, and how and where we build.

Given its strategic importance, Norfolk will ultimately be able to call on more resources to protect itself than most other US cities, not to mention those in developing countries. Yet, its travails demonstrate that even the most powerful places in the wealthiest countries will have to contend with a warming world and will not be spared from having to make tough choices. Lessons on how to deal with these choices successfully are needed now more than ever. Because climate change will present us with unprecedented risks, developed and developing countries alike are learning together. In fact, many poor countries are innovating faster because they have less time to spare and less margin for error. This book describes lessons communities have already learned to cope with climate impacts, as well as promising policy ideas for the years and decades ahead.

THE TIME FOR RESILIENCE IS NOW

For the first time in our brief history on this planet, humans have become a force capable of altering fundamental aspects of how the planet works, including its climate system. Our emitting of carbon dioxide and other heat-trapping gases by, among other things, burning coal, oil, and other fossil fuels on a global scale, has begun to disrupt the extraordinarily hospitable conditions that have allowed human civilization to flourish. Experts have given this new era a name, the Anthropocene, an era in which humans have become a dominant force in the world's natural environment. A central feature of the new era is accelerating climate change fueled by human activity.

Once regarded as a threat in the distant future, the impacts of climate change are now daily news stories. Extreme heat and heavy precipitation events are happening with greater regularity than in the past. Wildfires are becoming more frequent, as are floods. Seas are warming, rising, and acidifying. Climate change-related impacts are harming the health of humans, plants, and animals, and damaging livelihoods and entire economies. These impacts are here to stay and may get worse over time, depending on the trajectory of our greenhouse-gas emissions.

Resilience is urgently needed to enable communities to cope with the climate impacts we are already experiencing, as well as with future impacts. We use the term "resilience" to refer to the capacity of a community to reduce, absorb, and recover from the impacts of climate change.[1] This definition covers a wide range of actions, from "no regret" measures that can be taken quickly, at little cost, and with little controversy, to large-scale transformational changes that require sophisticated scenario and cost-benefit analysis, extensive consultations with the affected communities, and years to implement. "No regret" actions include things like the deployment of early warning systems for natural disasters and disease, or

1. In this book, we sidestep a long-running definitional debate about the differences between climate "resilience" and "adaptation." The terms are sometimes used interchangeably in the literature. Other times, they are assigned different definitions that get more and more complicated over time. We find this debate to be largely academic and not particularly helpful to business and government decision-makers whose main preoccupation is to manage climate impacts effectively. For that reason, we use only the term "resilience" in this book.

climate change education for engineers, lawyers, and health care professionals. Transformational measures include changing the shape of cities by making different land-use choices, significantly altering crop composition and agricultural practices for millions of people, and relocating entire communities.

Building resilience is not a substitute for reducing greenhouse-gas emissions. Indeed, cutting emissions is the best resilience strategy of all because it can safely spare us from some of the worst impacts of climate change. Another approach sometimes discussed for dealing with climate change impacts is geoengineering— intervening directly in the earth's natural systems by, for example, injecting particles into the atmosphere to block out the sun's radiation. But this kind of geoengineering remains a desperate move that carries frighteningly large risks and uncertain benefits.

Resilience will not magically spare us from the wrenching disruptions that accompany a warming world. It only acts as a shock absorber that can reduce the blows without entirely neutralizing the impacts. Resilience can help blunt some of the worst impacts and save lives, as well as protect the most vulnerable in society. It can enable some communities to find new ways to live and reach a new balance in a transformed environment. And in places where no such balance is possible, resilience means enabling people to settle in places where it may be possible.

The urgency of resilience is growing because we are not making progress fast enough in reducing greenhouse-gas emissions. The 2015 Paris Agreement provided a global framework for cutting emissions, and many countries are engaged in vigorous efforts to do so. But though the global emissions curve has at times shown signs of flattening, it has not bent decisively downward, and as of this writing, it has regained its upward tilt.[3] The risk that we will fail to cut emissions fast enough makes resilience all the more urgent.

If we breach the global average temperature thresholds that scientists are warning us about—2°C (3.6°F) and, increasingly, 1.5°C (2.7°F) above preindustrial temperatures—resilience measures will be vital to buffer the ensuing disruption. If we don't start investing in resilience now, we will still have to adjust to a changing climate, but the process will be considerably more expensive, chaotic, and traumatic. The demands of coping with a new and unfamiliar climate, and simply dealing with the growing exigencies of the present, will come to occupy more and more of our time and energy. Eventually, the privilege of great civilizations—dreaming about and shaping the future—will no longer be ours.

HOW WE CAME TO WRITE THIS BOOK

Like Ann Phillips and Christine Morris, we came to climate resilience from unconventional places. Alice spent the bulk of her career in courtrooms, first as a US federal prosecutor chasing white-collar crooks and, later, as a judge presiding over cases ranging from homicide to medical malpractice. She became immersed in climate change after she joined the US Department of Homeland Security in 2009 as senior counselor to the secretary. She was tasked with helping the department understand how climate could affect its operations. This was not immediately obvious. What did climate change have to do with an agency in charge of policing US borders, providing emergency relief, enforcing immigration policy, protecting infrastructure, and stopping terrorists from boarding planes? So Alice set up a task force and, before long, discovered that climate change impacts will touch virtually all the department's work.

Alice went on to the White House to lead resilience efforts as special assistant to President Obama and member of his climate team. There, she oversaw the development of a national flood risk-management standard and led efforts to establish additional risk-management standards for other natural hazards. She also oversaw the development of executive orders on managing national security risks from climate change, making international development assistance "climate smart," and improving federal efforts to prepare for a warming Arctic.

Leonardo spent several years as an academic, writing about international cooperation in the areas of finance and economic development, before encountering climate change as an all-absorbing preoccupation. After he joined the Obama Administration, one of his first tasks as head of the policy office at the US Agency for International Development (USAID) was overseeing the development of a climate change strategy for the agency.

Later, as deputy assistant secretary for Energy and Environment at the Treasury Department, he negotiated finance elements of the Paris Agreement and represented the United States on the governing bodies of major providers of climate finance, including the Green Climate Fund and the Global Environment Facility. He also led initiatives to enable developing countries to access disaster risk insurance, especially against climate-related hazards. Having as a child experienced a catastrophic earthquake that destroyed parts of his hometown of Mexico City, in 1985, the issue of building resilience against climate change resonates deeply with him. Communities can and must learn from disaster.

WHAT THIS BOOK OFFERS

This book is intended primarily for those interested in the practical aspects of how to prepare for the coming climate disruption. Our intended audience includes current and aspiring decision-makers in the private sector, civil society, and all levels of government, as well as young people today who are considering which career path to pursue. It also encompasses readers who simply want to deepen their understanding of the risks posed by climate change and the choices the government and private companies can make now to build resilience and manage those risks. Although we focus primarily on US examples and insights, we also draw on the experiences of other countries, including Bangladesh, China, Colombia, France, Fiji, India, Mexico, the Netherlands, and South Africa. We have tried hard to write a book that is rooted in the facts, but does not dwell on catastrophe or wallow in apocalyptic visions of disaster. Fear, helplessness, and paralysis are the last things that decision-makers and the American public need right now. Instead, we have sought to be pragmatic, to provide actionable strategies, and to highlight the promising efforts that are already under way.

Our work draws on our personal experiences in the federal and state government, as well as on the stories of mayors, engineers, city managers, bankers, generals, health care providers, lawyers, economists, and entrepreneurs who are working to build climate resilience. Our goal is to put these stories in a larger context, so that they are not only engaging but will also be useful to those who want to look beyond today's polarized debate on climate change. We also wanted to write a book that breaks with the current debate in which valuable insights about resilience typically remain trapped inside professional and academic disciplines, isolated from each other by technical language and specialized communities of practice that mostly talk only to themselves.

We offer ten lessons for building resilience, organized in ten chapters. These lessons do not attempt to cover the exploding field of climate resilience comprehensively. We focus on what we know best, and as a result, the built environment gets plenty of attention, as compared to, say, fresh water access, ocean health, or agriculture. At the end of each chapter, we provide a short list of prescriptions and provocations to further stimulate debate.

SYSTEMS FOR LARGE-SCALE CHANGE

The first three chapters are about systems that can drive resilience on a large scale. If we can get these systems to work for resilience, we can strengthen relatively quickly the capacity of whole cities, states, and even countries to cope with the impacts of climate change. To be sure, changing these systems takes time, dedication, and political will, but it is vital if we are going to meet the climate challenge. Building resilience gradually with piecemeal efforts simply will take too long and leave communities exposed to unacceptably large risks.

One system involves the rules and practices that govern where and how we construct housing, commercial and industrial buildings, and infrastructure. This chapter observes that communities tend to overhaul their building rules and practices only after they experience extraordinarily traumatic shocks, after they live through what we call "no more" moments. But we can avoid human suffering and enormous expense if we overcome this tendency and strengthen our building system before the "no more" moment arrives.

Another system with potential for large-scale impact is the judicial system. In the United States, as in many other places, the courts move slowly. But in the US legal system, which is based on precedent, a single case can have a disproportionately large impact, as governments, companies, and individuals change their behavior to avoid legal liability. In this chapter, we look at evolving legal thinking about who should pay for climate change damages and who should be held responsible for failing to anticipate and plan for climate risks. As in other areas of social change, the law could eventually help drive large-scale resilience.

The third system consists of markets in which various kinds of assets are traded. We look at four different markets—the markets for stocks, bonds, property, and insurance. If the prices for these assets change in response to accelerating climate risk, then thousands of investors, homeowners, and insurers should respond by changing behavior. We examine how transparency is working in each of these markets and conclude that while enhancing disclosure of climate risk offers an alluring "theory of change," the approach encounters several pitfalls.

TOOLS FOR THE DECISION-MAKER

Next, we group three lessons under the umbrella of tools that leaders in business and government can deploy to build resilience. Money—both to prepare for climate change impacts before they arrive and to recover and rebuild after they do—is, of course, essential. Yet, governments at all levels are struggling to raise money for resilience. In this chapter, we look at different techniques that communities are adopting to finance these large expenses. Some, like taxation, international assistance, and reinsurance are old-fashioned but effective. Others, like value capture, carbon pricing, cat bonds, and national reserve funds, range from the relatively new to the entirely experimental.

Climate data and information also provide critical instruments for sound decision-making on resilience. This chapter describes a paradox we currently face. Thanks to technology, we are living in the golden age of climate and weather data; we enjoy more extensive data collection than ever before and immensely powerful analytic tools, including artificial intelligence. Yet, many of the people who need this information most can't access it, or they can't apply it to solve concrete problems because the information is not being translated into a form that users can understand. Getting the data—and making them usable—is a central lesson to build a resilient tomorrow.

Another tool available to decision-makers comes from the field of behavioral economics. Cognitive biases can make it difficult for human beings to tackle complex problems, including climate resilience. Behavioral scientists tell us that because human nature is hard to change, it is best to work with it. In this chapter, we identify techniques that try to "nudge" people toward resilient behavior, be it reducing their water consumption, planning for an uncertain future, or taking disaster evacuation orders seriously.

THE UPENDERS

We devote the final section of the book to the toughest lessons. These involve some of the most daunting challenges that climate change will hurl at us, challenges we are still grappling to understand. These challenges will upend many aspects of human life—human health, economic and social inequality, migration patterns, and geopolitics and national security. Confronting these issues will first demand that we re-examine fundamental assumptions about how the world works, including the assumption that our past experiences will remain a sound guide to the future. It will also demand skillful political communication, as our leaders will need strategies to engage the public on these difficult issues.

One challenge involves the escalating threats that climate change is presenting to human health. Vector-borne diseases are likely to spread, and rising temperatures and extreme weather will affect many different dimensions of human health, including mental health. Climate impacts will also complicate the delivery of health care. We focus this chapter on how to protect the reliability and effectiveness of the health care system in a warming world. Facilities must be hardened against extremes, but the system must also become smarter, adopting better predictive analytics, disseminating preventive information faster, and training health care professionals about climate change.

Climate change is also magnifying existing economic and social inequality, disproportionately affecting the poorest and most vulnerable across and within countries. Warming temperatures will affect geographies and social groups unevenly. Perversely, the communities likely to suffer the most climate-induced economic declines are already among the poorest. Resilience to climate impacts will depend crucially on access to resources like private savings, insurance, transportation, and social connectedness. But access to these things is already highly unequal in many areas. Thinking creatively about how to protect the poorest and most vulnerable will be essential to buffering growing inequality.

Unchecked climate change will lead to mass migration in some parts of the world. This chapter finds that while it is often best to help communities remain in place, some people will have to move in response to changing environmental conditions. We can wait until the displacement is sudden, chaotic, and traumatic, or we can spare people much hardship by planning ahead. We can strengthen receiver communities, encourage people to move preemptively, and discourage people from staying in dangerous places before climate impacts drive them out. Leaders will need to develop a persuasive political strategy to engage the public.

Finally, in the last chapter, we describe how climate change acts as a "great disrupter" in international relations. Its impacts will likely upend traditional assumptions about national security, creating new power vacuums where bad actors can take root, threatening power bases previously thought to be invulnerable, and intensifying resource competition among countries. We borrow a program of action developed during our time in government to offer potential ways forward on this issue.

Our hope is that these ten lessons, and the stories that underpin them, will offer constructive ideas for those interested in building a resilient future, including today's and tomorrow's decision-makers in the United States and beyond. We predicated the lessons upon the idea that, even if we cannot avoid some of the impacts of climate change altogether, the resilience we build today can spare millions of people from needless hardship. What we seek, above all, is both an antidote to despair and a cure for complacency.

PART I

SYSTEMS FOR LARGE-SCALE CHANGE

[1]

RETHINK WHERE AND
HOW WE BUILD

Hurricane Andrew sliced like a scythe across the state of Florida in August 1992. Driven by sustained winds over 150 miles per hour (240 km/h), the storm surge towered 17 feet (5.2 m) above Biscayne Bay, near Florida's southern tip.[1] Andrew destroyed 25,000 homes and left many more damaged. All told, it caused $25 billion in damage in what was then the costliest disaster in US history.[2] After Andrew subsided, officials in Dade County convened an investigatory grand jury to determine how the storm could have caused such extensive damage. The investigation exposed lax enforcement of building codes and shoddy construction practices. Less than two years after the publication of the scathing report, Florida had not only adopted the toughest wind building-code standards in the nation but had begun enforcing them.[3] "Andrew's most obvious lesson was that we were not prepared for this hurricane, neither as individuals nor as a community," the grand jury's final report read. "This mistake must not reoccur."[4] Florida, it seemed, was crying "no more."

Communities tend to learn things the hard way, reacting in the wake of disasters rather than in anticipation of them. Once a truly devastating extreme event hits, we swear "no more," and we build back better—at least, well enough to prevent another similar disaster from causing as much damage. But if we prepare ourselves in advance, before a disaster strikes, we could save lives, time, and money. Preparing in advance carries particularly large rewards when it comes to making investments in the built environment—our houses, schools, businesses, ports, roads, wastewater treatment plants, and electric grids. Building infrastructure costs a lot of money. We spend heavily on it because we intend these buildings and systems to last thirty, forty, or fifty years or more. And with climate change impacts already causing ever-greater destruction, communities need to start rethinking the rules and practices that govern how they build.

Virtually all our existing infrastructure was designed to withstand the extremes that have been experienced in the past. Because, historically, scientists—let alone city planners, engineers, architects, and builders—could not project the impacts of climate change with much precision, our existing design choices and plans for infrastructure have largely ignored the risks posed by those impacts. In fact, building plans continue to ignore climate risks in most instances. This means that, today, even major cities have found themselves regretting their failure to account for sea-level rise, for example. Shortly after the San Francisco–Oakland Bay Bridge reopened in 2013 after extensive reconstruction, the new entry ramps needed flood remediation to deal with sea-level rise. On the other side of the country, New York City's brand-new South Ferry subway station—built at a cost of close to a half billion dollars—flooded during Hurricane Sandy, turning into "a giant fish tank."[5]

This chapter identifies strategies that communities and individuals can take now to strengthen their building practices to endure new extremes driven by a changing climate. Among other things, it analyzes how improving building codes and standards and insisting on wiser land-use policies, especially in the absence of a "no more" moment, can serve as a bulwark against the destruction that climate-fueled disasters bring. Making better decisions about where and how we build today can protect communities against a "no more" moment tomorrow.

LEARNING WITHOUT THE "NO MORE" MOMENTS

On the night of January 31, 1953, an unusually high tide, coupled with a severe windstorm blustering off the Netherlands's northwest coast, overwhelmed scores of the country's dikes and dams. The flood waters rushed in, surging as high as 15 feet (4.5 m) in some places, killing over 1,800 people, and forcing tens of thousands of residents to evacuate. The sea water spilled across farmlands, rendering them useless for years. To this day, on February 1 each year, the Dutch commemorate those who died in the *Watersnoodramp*, or the "water emergency disaster" of 1953. This was their "no more" moment.

Once the waters receded, the Netherlands realized that simply trying to keep water out with dikes, levees, and canals could not protect the country. Instead of engaging in an ad hoc struggle against constant flooding in a nation where about a quarter of all land lies below sea level, the Dutch began to think bigger, reimagining how they could live with the water instead of fighting it. Today, the Netherlands has some of the most sophisticated flood measures in the world.[6] With outdoor parks designed as back-up reservoirs for flood waters and floating neighborhoods, the Dutch have invested deeply in flood mitigation. As the mayor of Rotterdam astutely observed in 2017, "We must learn to live with water . . . That's just common sense."[7]

After "no more" moments in the Netherlands and Florida, the political will to undertake large-scale and decisive action coalesced. Governments acted. Money was raised and spent. Legislation was adopted. Rules were finally enforced. These crises proved so severe, so at odds with common experience, and so traumatic to the affected communities that a collective decision was made to do whatever it would take to ensure "no more." But just because disasters can spur change, it doesn't mean they always do. In fact, only extreme catastrophes seem to lead to decisive action. The rest of the time we largely fail to learn from our experience. The reasons for inaction are many: increased construction costs, potential loss of tax revenue, fear of litigation, optimism bias (as we will see in chapter 6), and plain old inertia.

The argument for acting before a "no more" moment happens is compelling. Buildings that have been designed and constructed according to strict codes and standards typically suffer less damage from extreme weather events. In 2005, the US National Institute of Building Sciences (NIBS) published a study that reviewed several thousand government-funded projects designed to reduce the risk of damage from floods, wind, and earthquakes. The study concluded that one dollar spent on these investments saves society an average of four dollars.[8] In 2017, the NIBS updated the study and found that investments in resilience generated an even greater benefit than the original estimate—six dollars saved in averted disaster costs for every dollar invested in resilience.[9] A pair of academics who analyzed two dozen disaster-mitigation studies from around the world similarly concluded that the median benefit-cost ratio was in the neighborhood of six to one.[10] And a NIBS study from 2018 concluded that adoption of the latest model building codes would generate a national benefit of eleven dollars for every dollar spent on building to code.[11] The national benefit results from a host of factors from lower property damages to fewer business interruptions.

Why, then, do we fail to learn without "no more" moments? Take the case of building codes and standards in the United States. The United States has no single, national building code. The decision to adopt building codes lies in the hands of state and local jurisdictions, and so a patchwork of building-code standards stretches across the country. States and localities usually model their building codes on templates developed by nonprofit organizations. Adopting all or part of a model code saves state and local governments the expense and trouble of creating their own.

A commonly used model building code is one that was developed by the International Code Council (ICC), a nonprofit organization whose membership includes representatives from government and the building industries. But though all fifty states have adopted some version of the ICC model building code, not all jurisdictions adopt the same version. Sometimes, political opposition from builders and developers prevents state and local lawmakers from adopting the most recent version of the ICC model building code or applying it evenly to all types of construction. According to the Insurance Institute for Business and Home Safety, a nonprofit building association supported by the insurance industry, eight of the eighteen most hurricane-prone states in the United Sates have no mandatory statewide code.[12] In addition, virtually none of the model codes take climate change into account. This flawed system has allowed many developers in hurricane-prone states to build homes without added reinforcement for gale-force winds or towering storm surges. Because building below code requirements typically costs less, builders can sell such homes at lower prices. By the time catastrophe strikes and the home gets damaged, the builders and realtors have walked away, cash in hand. To make matters worse, the model building codes do not yet account for the future risk from climate-exacerbated extremes. The question then becomes: Who pays for the damage? The homeowner? The insurance company? Taxpayers, through government bailouts?

THE HIGH COST OF CHEAP CONSTRUCTION

An often-cited excuse for not strengthening building codes is the high cost of improvements, but in the long run, the costs of building relatively cheap but vulnerable infrastructure often dwarf the benefits. Consider the case of Houston. After Hurricane Sandy spurred Congress to approve more than fifty billion dollars in disaster aid in 2012, President Obama took a stab at reducing the financial risk to the federal government posed by natural disasters. In 2017, he signed an executive order creating a federal flood standard. It applied to structures in or near floodplains built or substantially improved with federal taxpayer money, and required that they be elevated by two or more feet (or in accordance with the latest climate science).[13] Every federal agency affected by the order agreed to it before it reached the president's desk for signature. President Trump, however, rescinded the order in August 2017, to the delight of some builders, since the Obama regulation raised the cost of construction. Ten days later, Hurricane Harvey hit Houston, reminding the nation why building standards matter.

Houston, the largest city in Texas and the fourth largest US city overall, historically sacrificed strict building codes in favor of creating cheap housing to promote rapid growth. The city covered its flat terrain with miles of pavement, failed to invest in proper stormwater-management systems, and issued permits allowing the construction of hundreds of homes in a dry reservoir that the US Army Corps of Engineers had designated for capturing flood waters. When Hurricane Harvey slammed into Houston in 2017, it let loose a record-setting 50 inches (127 cm) of rain on the hard clay soil and miles of pavement. Hundreds of thousands of homes were flooded and almost a hundred people lost their lives. The total economic damage climbed to more than $125 billion.

Perhaps Harvey was Houston's "no more" moment. Just seven months after Harvey struck, the Houston City Council adopted requirements to elevate any new home in the 500-year floodplain (the area that has a 0.2 percent chance of flooding in any given year) by two feet (0.6 m). The city estimated that if the regulation had been in place when Harvey struck, 84 percent of the structures flooded by Harvey would have avoided that fate.[14] We will only know if Harvey inspired real change once Houston actually enforces the tough new building codes it adopted after the disaster. But what's clear is that the city could have spared itself, its citizens, and the federal government significant costs had it chosen to invest in resilience earlier.

Norfolk, Virginia, one of the cities most threatened by sea-level rise in the United States, is trying to learn without a "no more" moment, despite the higher construction costs. In 2018, the city rewrote its zoning ordinance for the first time in over a quarter-century. The new ordinance increased its flood-protection requirements, so that even buildings outside areas of high flood risk have to be elevated. The added cost of elevation concerned some developers and real estate agents, as well as city council members. One local builders' association estimated that the price tag on a single-family residence could jump by close to $15,000 because of the new rules. Yet despite the resistance, the city forged ahead to insist on greater flood resilience. As in Houston, the question is whether Norfolk will actually enforce the new ordinance.

In some cases, improvements that render infrastructure resilient do not always carry additional cost. Consider the case of wildfire protection. A recent study disproves the traditional assumption that building a code-compliant home is more expensive than building a less-resilient home. Switching exterior paneling from (flammable) cedar planks to fiber cement saves so much money that it offsets the extra cost of building the rest of a code-compliant home. In some regions, it may even cost less to construct a wildfire-resistant home than a conventional home.[15]

BUILD FOR THE FUTURE, NOT JUST THE PAST

For virtually all existing infrastructure, designers and developers have relied on historical events to determine the extremes that infrastructure investments will need to withstand during their service life. Infrastructure, including roads, ports, wastewater treatment plants, railways, bridges, dams, tunnels, power plants, and airports, is typically designed to handle the types of extreme weather events that have occurred in the past. It has not, in all likelihood, been built to withstand the strongest storms that may occur in the next fifty or one hundred years in a warming world. Nor have many designers and developers considered the increased operational expenses that climate-fueled extreme events may bring. You might expect that this would not be true for some of the most zealously protected infrastructure of any society, the kind used for military purposes. But you'd be wrong.

Kwajalein Atoll is a tiny island in a chain of over a hundred small spits of land in the South Pacific. Kwajalein is home to the US Army's Ronald Reagan Ballistic Missile Defense Test Site, which, among other things, is used to test interceptor missiles that can shoot down intercontinental ballistic missiles aimed at the United States' western coast. In 2014, the US Air Force awarded Lockheed Martin a $915 million contract to build a radar installation on the island to replace a decades-old system.[16] The new system, named "Space Fence," dramatically improves the ability of the Air Force to track objects in outer space as small as a baseball and to maneuver spacecraft and satellites out of the way to avoid disastrous collisions.

Before construction began, the US military had dismissed concerns about sea-level rise in Kwajalein and the troubles it would cause for the more than one thousand people who live and work there. The Department of Defense's initial risk assessment concluded: "Based on historical data, there are no anticipated issues with ocean tide and/or wave flooding."[17] But four years later, after construction was underway, the department commissioned another study of Kwajalein to explore future risks. It found that flooding from sea-level rise threatens Kwajalein's fresh water supplies in the very near future, and that the majority of the atoll could flood annually by 2055—almost a dozen years before the United States' lease on the island officially ends.[18]

Critical civilian infrastructure is at risk, too. Take the Gulf Coast of the United States. Four feet (1.2 m) of sea-level rise in the region (which is well below the upper extreme of eight feet (2.4 m), now projected by the federal government) would flood one-quarter of the major roads, 10 percent of the region's rail miles, 75 percent of all freight facilities, and at least one international airport.[19] In 2015, *National Geographic* reported that rising sea levels put thirteen nuclear power reactors at risk.[20] Dozens more are at risk of inundation due to dam failure, which can be hastened by extreme precipitation.[21] And it's not just water that threatens nuclear power plants—so do wildfires and warmer temperatures.

Designing, building, operating, upgrading, and eventually retiring infrastructure projects is an expensive process. As aging infrastructure is repaired and retrofitted, and as new infrastructure is constructed, incorporating future risks from climate change impacts into their design, construction, and operation must become routine. Failing to take those risks into account will likely lead to increased maintenance and operating costs and shortened service life. Failing to screen projects for climate resilience will lead to more waste and more infrastructure collapse. The stronger infrastructure required may come with a significant upfront cost. But employing thoughtful design strategies can allow for making relatively modest resilience investments now, while still planning for more extreme events down the road. This approach could well be cheaper than having to retrofit the infrastructure from scratch if conditions worsen beyond expectations.

Part of the challenge is that architects and engineers are not being systematically trained to think about how climate change will affect what they build. Few understand this better than Daniel Kreeger, an information-technology and communications entrepreneur from Florida who has turned himself into a tireless champion for climate education. Kreeger noticed that in the United States, architecture firms do not require that architects possess the skills necessary to design and account for the coming climate extremes. So he visited the American Institute of Architects, the credentialing body for the architecture profession in the United States, and tried to persuade its officials to include climate change-related material in the credentialing requirements for new architects.

Institute officials replied they could include the material in their code of conduct, but said they could not make it a requirement because neither the market nor the law currently demands it. "We have a chicken-and-egg problem," Kreeger explains. "Employers aren't requiring it, and the suppliers—the credentialing bodies and universities—can't make it the standard."[22] In the future, liability laws may break the circle (see chapter 2). But hopefully, the market will start demanding these skills, pushing the credentialing bodies to make them a requirement, before fears of legal liability make it imperative.

"LOW-REGRET" DECISION-MAKING

Nobody knows for sure the precise stresses that a piece of infrastructure will have to endure in a changing climate. Given the current state of climate modeling, the nature of expected future impacts remains uncertain. Yet decisions about design and construction must be made every day. Confronted with such uncertainty, "low-regret" decision-making is particularly important for decision-makers. It involves accounting for the best current estimates of future conditions while designing flexible structures to allow for modification if circumstances worsen. Low-regret decision-making advances communities down the path toward resilience but gives them enough wiggle room to adjust as conditions and needs change. Take, for example, improvements made to the Los Angeles–San Diego–San Luis Obispo Rail Corridor, the second-busiest intercity passenger rail corridor in the United States. To account for potential future sea-level rise, the rail authority invested in pylons strong enough to accommodate the future elevation of a bridge should flood conditions worsen over time.[23]

Similarly, the Dutch have demonstrated how seawall design can benefit from flexibility. They have raised dikes to account for the most likely amount of sea-level rise. If the seas rise higher, the foundations of the dikes have sufficient strength to accommodate subsequent retrofits that will increase the height of the dikes.[24] Similarly, the US Naval Academy in Annapolis, Maryland, announced that it will raise one of its seawalls up to 3 feet (1 m) in three decades. The wall's design can accommodate an additional increase of 1.6 feet of height (0.5 m), allowing for uncertainty in the projections. Yet another example is the continual updating of municipal building codes. By regularly incorporating the latest climate science into building standards, as the Obama flood standard did, communities will slowly build resilience when structures undergo retrofits and rebuilds. Such a process gradually increases the resilience of a municipality's building stock as time goes on, even when the future remains uncertain.

As we have seen, the issue of cost will typically loom large in most building decisions. To promote low-regret investment in flexible design, governments can encourage communities to make marginal additional investments that enable future design modifications as conditions change. One potential strategy is for the government to provide financing at the beginning of a major infrastructure project to cover the cost of the flexible features. That the government should cover these costs makes sense. Since it's often the public purse that ends up footing the bill when underprepared communities face devastation, that flexibility will save the government money in the long run.

WHERE WE BUILD MATTERS TOO

As climate change advances, we must also fundamentally reconsider where we build, not just how we build. Building in inherently high-risk areas leaves people's homes and lives vulnerable. The severity of the risks posed by more extreme storms, bigger wildfires, and higher sea levels requires that "no more" must truly mean "no more" for some locales: people will have to move out of harm's way. We cover this in greater length in chapter 9, on relocation.

Australia's "no more" moment came when some four hundred brushfires ignited in the state of Victoria on a single Saturday in 2009. The so-called Black Saturday fires started on the hottest day recorded in Victoria since record-keeping began. They destroyed thousands of homes, killed almost two hundred people, and burned over a million acres (400,000 hectares). In the aftermath of the fires' destruction, Australia fast-tracked new building regulations for bushfire-prone areas. Standards now require a risk assessment for all properties, including those outside wildfires areas, and builders in high-risk areas must obtain a special permit for new construction. But then Australia took it a step further. The Victorian government also pursued a "retreat and resettlement" strategy by creating a voluntary buyback program allowing the government to purchase land with an unacceptably high risk of fire.[25] In that moment, Australia prepared proactively for increasing wildfire risk by reconsidering not just how, but also where to build.

Worldwide, soaring temperatures and increasingly prolonged droughts have contributed to hotter, drier conditions that exacerbate the risk of wildfires. But as wildfire risk has grown, so has the number of Americans choosing to live in wildfire-prone areas, building homes in what is known as the *wildland-urban interface,* or WUI (pronounced *woo-ee*). Between 1990 and 2010, Americans built in the wildlands at a rate of 5,500 houses per day, converting them into WUI at the remarkable pace of 4,000 acres daily. According to the 2012 Census, 120 million people—more than a third of the US population—now call the WUI home.

Between 2017 and 2018 alone, California suffered seven of the twenty most destructive fires in its history, not including the largest wildfire ever recorded in the state, the Mendocino Complex Fire in Northern California, which burned an area almost the size of Los Angeles. In 2018, the Camp Fire was the state's deadliest and most destructive in its history, and the world's costliest disaster that year. One-third of Californians already live within the WUI but, so far, the state has struggled to muster the political will to curb development in the highest-risk areas.[26]

Take, for example, a proposal to permit the development of a five-mansion compound on a fire-prone hillside in the city of Malibu in 2015. The federal government advised against it, pointedly warning that "the placement of homes on a ridgeline documented to have burned at least six times between 1942 and 2010 makes it almost certain the ridgeline will burn again in the near future."[27] But the project was approved anyway. And even after the historic 2018 wildfires, Los Angeles County supervisors approved a 19,000-home community to be built in rugged mountains 65 miles (105 km) north of downtown Los Angeles, in an area already designated as a "high" and "very high" fire-hazard zone.[28] A key question that the supervisors will need to ask themselves when those homes go up in smoke is: Who should pay?

SELF-HELP RESILIENCE

As governments dawdle, enterprising citizens are exploring how they can build resilience on their own. Mark and Valerie Sigler are two such citizens. The Siglers had lived in Pensacola Beach, Florida, for fifteen years before they said, "no more." Storms tore at their home year after year, and after the Siglers had rebuilt it time and time again, they began to think of a better way. The Siglers dreamed of a home that could survive the punishing combination of wind and rain inflicted by the hurricanes that regularly storm through the Gulf Coast. They designed and built what they endearingly named the Dome of a Home (Figure 1).

To achieve their dream of a resilient home, the couple "researched, dreamed, politicked, and sweated blood, sweat, and tears to get it built," recalled Valerie Sigler. The shape of the house allows water to flow around it and prevents pressure from building up along the structure. The building also contains many subtle

Figure 1 The Dome of a Home, Pensacola Beach, Florida.

design choices that make the house specially adapted for hurricane conditions, such as the breakaway front steps that detach from the home to avoid damaging the rest of the structure.

Finished in 2003, the Dome of a Home faced its first major challenge, Hurricane Ivan, in 2004. As news crews flocked to Florida to cover the storm, word about a home that could withstand a hurricane spread. Suddenly, MSNBC, Discovery, National Geographic, NBC, *Good Morning America*, the *Today Show*, the Travel Channel, and others all wanted a glimpse of the Siglers' Dome of a Home. Mark Sigler even chose to wait out the raging storm inside the house with an MSNBC news crew.[29] Ivan destroyed neighboring homes, but the dome remained largely intact. As Valerie put it, Ivan left the home with "no structural damage, just cosmetic . . . exactly as engineered."

The house was not cheap. The Siglers did obtain a Federal Emergency Management Administration (FEMA) grant to cover over $200,000 of its cost, but their 6,000-square-foot home (560 m²) likely cost upward of $750,000 to build.[30] Although the Dome of a Home was a valiant effort to confront hurricanes, its high cost makes it impractical as an easily replicable solution, especially for low-income housing. However, some of the home's resilient-design features could be incorporated into lower-cost structures.

Other private citizens and companies have similarly invested in developing resilient designs. Morphosis Architects, an innovative architecture firm based in New York City and Los Angeles, created a prefabricated flood-proof home they have dubbed the Float House, designed to survive Hurricane Katrina–like flooding.[31] Its foundation acts as a raft so that the house rises with the water level. Another prefabricated house, constructed in Stinson Beach, California, has a "recreation area" on the ground floor designed to break free and float away to keep the rest of the elevated house from sustaining damage.[32] In Hanoi, a Vietnamese firm has created the Blooming Bamboo Home, which floats to survive floods up to 10 feet (3 m).[33] And the ever-resourceful Dutch, with government help, are building a floating neighborhood in a particularly flood-prone area of Amsterdam.

RECRUITING NATURE

So-called *gray infrastructure* projects, such as seawalls, levees, and breakwaters, are often the solution of choice when it comes to building resilience to sea-level rise and increased flooding. In Norfolk, in 2014, government leaders met with local citizens to discuss the challenge to the area from accelerating sea-level rise. At the meeting, one local resident, unsurprisingly, cried, "Let's just build a seawall!" It's a simple solution. Norfolk had already installed one seawall in 1970, and it has plans to build others. But relying solely on gray infrastructure could prove to be a costly mistake in instances when green infrastructure could provide better protection at a cheaper price than building with concrete.

Green infrastructure, such as mangroves, wetlands, and coral reefs, can buffer storm surge, and vegetated areas can absorb excess precipitation. Green infrastructure can also blunt damage from extreme weather events. In 2016, experts estimated the difference coastal wetlands made in protecting the northeastern coast of the United States from Hurricane Sandy. Researchers compared the actual damages against a model that projected what would have happened if the region had no wetlands at all. Ultimately, the researchers found that the "risk reduction ecosystem services" of wetlands prevented about $625 million in damages regionally.[34]

Green infrastructure sometimes performs much of the work of gray infrastructure, but more cheaply. In 2007, the US Environmental

Protection Agency reviewed thirteen real-world cases of stormwater-management projects in communities that had considered both gray and green infrastructure projects. In eleven of those cases, green infrastructure strategies reduced costs by 15 percent to 80 percent compared to conventional (gray) infrastructure strategies.[35]

In view of these benefits, China has moved decisively to advance flood mitigation, seeking to reverse the trend toward increased flooding in urban areas. Green infrastructure is a consistent theme in this effort. In 2014, President Xi Jinping launched an ambitious stormwater-reuse program to address the problem of recurring urban flooding, aptly named the Sponge City Initiative. The program's long-term goal is to achieve 70 percent reuse of stormwater runoff in 80 percent of China's urban areas by 2030.[36] Kongjian Yu, a Chinese landscape architect who was instrumental in developing the Sponge City program, warns: "It's important to make friends with water."[37] One approach to reducing flood risk in a Sponge City involves preserving and restoring natural waterways and installing other measures that mimic natural systems, such as green roofs and pavement that allows water to pass into the ground below.[38] In 2017, Berlin followed China's lead and launched its own Sponge City initiative to tackle increased flooding and heat extremes.

But despite the benefits of green infrastructure, these investments all too often fail to attract serious consideration. Governments are more accustomed to funding construction workers and traditional engineering firms than they are to funding mangrove caretakers and urban forestry stewards. But if communities truly want to build resilience to current and future climate change without compromising their fiscal health, then pondering the protections afforded by the natural environment is a smart place to start.

Throughout history, we have seen that communities can learn resilience in the wake of truly catastrophic events. But we must learn to act before the "no more" moment strikes. Doing this will spare lives and save billions of dollars. When planning where and how to build, we can no longer rely solely on past events to guide our decisions. We must also start to recognize that investing in cheap construction that is not resilient will often carry bigger costs when climate impacts strike. Instead, we must account for the fact that climate change will give rise to a new environment, and the rules and practices that govern our building choices must change now to build a resilient tomorrow.

PRESCRIPTIONS AND PROVOCATIONS

- Federal, state, and local governments should screen all government-funded infrastructure projects for climate risk and require measures to address anticipated climate risks during the service life of the project.
- Federal, state, and local governments should require, as a condition of public investment in an infrastructure project, that the design options consider green infrastructure, as well as low-regret strategies that offer greater flexibility in the face of climate uncertainty. The federal government should consider providing upfront financing to help cover the costs of the flexible features of low-regret strategies.
- Organizations that develop model building codes should immediately accelerate and prioritize the development of climate-resilient model codes.

- Professional licensing bodies should establish core competencies and licensing requirements for architects, engineers, and others involved in the construction industry that require proficiency in designing and building for a changing climate.
- The federal government should restore the federal flood risk-management standard.

[2]

LAWYER UP

When a group of senior officials from the Department of Homeland Security visited Utqiagvik, Alaska, in 2012, the mayor hosted a dinner at Pepe's North of the Border, which surely served some of the best Mexican food in the Arctic.[1] Edward Itta was the long-time mayor of Utqiagvik, the northernmost city in the United States and home to the Iñupiat indigenous community. As the evening drew to a close, Mayor Itta shared his memories of what life used to be like in Utqiagvik. He described in poetic terms the changes that had come to the land where he grew up. The winter sea ice that had once formed along the coastline had grown thinner, and it disappeared earlier and earlier each spring. Ice cellars that had long ago been dug deep into the permafrost to store whale meat had begun to thaw, and the grasses moved farther north each summer. Itta lamented the vast changes his community had already experienced and expressed worry about the changes the warming temperatures might yet bring. Rising temperatures have contributed to the altered landscape Mayor Itta described. Indeed, the Arctic is warming twice as fast as the rest of the world.

Given the wrenching shifts already occurring in Alaska, it's not surprising that the state gave rise to one of the first lawsuits filed in the United States over climate-related damages. The village of

Kivalina, population four hundred, lies south of Utqiagvik and occupies some four square miles (10 km²) on the tip of a barrier island off the west coast of Alaska. The village is so small that it doesn't appear on many maps of the state, much less of the country. In 2008, the town of Kivalina filed a lawsuit against two dozen oil companies in federal court. Residents claimed that by contributing to greenhouse-gas emissions, the fossil fuel giants had added to the impacts of climate change that were now threatening the village— namely, the loss of sea ice and increased storms and flooding.[2]

Central to Kivalina's lawsuit was the question of who will pay for the damage and disruption caused by climate change. Indeed, the core issue in most lawsuits dealing with climate change damage boils down to the question: Who pays, and how much? These lawsuits seek not only monetary compensation for lost homes and endangered livelihoods, but also compensation to cover the cost of building resilience against the future impacts of climate change.[3] In the absence of government action to address climate change, either by a legislative body or a government agency, the answer to the question of who pays lies with the courts. Neighbors, children, homeowners, insurance companies, and others have turned, and will continue to turn, to the judicial system to extract an answer to that fundamental question.

The quest for an answer can lead to more than just dollar payouts—it can push government policy and private action toward greater resilience. In legal systems that rely on "precedent," such as that of the United States, judges use principles established in earlier cases to decide subsequent cases. A lawsuit that breaks new ground can set a precedent that may not only determine the outcomes of subsequent lawsuits but also influence the behavior of those seeking to avoid being found liable in the future. This chapter explores how litigation over the harm caused by the impacts of climate change could offer greater clarity on who should pay damages and thereby spur decisions to invest in resilience on a large scale.

FORCING THE GOVERNMENT TO ACT

When it comes to climate change impacts, governments have already found themselves sitting at the defense table in the courtroom defending a climate-related lawsuit. Those seeking to sue governments face barriers because governments enjoy legal immunity. But if an exception to governmental immunity can be found, lawsuits against governments could drive some of the most significant resilience efforts.

The climate lawsuits against governments that have garnered the most media attention have sought to force action to cut greenhouse-gas emissions. Brought by children, farmers, and others, the lawsuits do not seek monetary damages; they seek immediate government action to protect against increasing harm from climate change. As of this writing, courts have generally declined to allow these lawsuits to go forward, concluding that policy issues of this magnitude are better left to legislative bodies than to the views of any one court.

That wall may have begun to crack in some jurisdictions. In 2015, Ashgar Leghari, a twenty-five-year-old law student and farmer in Pakistan, sued the Pakistani government to force action on climate change. The appellate court found in Leghari's favor and ordered the government to increase its focus on the issue, even appointing a commission to monitor the government's progress. In 2018, a Dutch appellate court upheld a lower-court order instructing the government to reduce its emissions by at least 25 percent. And in 2015, eight kids and the climate scientist James Hansen sued the US government in federal court. In *Juliana v. United States*, the kids asserted that the government had known about the destabilizing impacts of climate change for over fifty years but had done nothing to stop carbon emissions from escalating to levels unprecedented in human history. Although the trial judge recognized that the case was "no ordinary lawsuit," she ruled that it could go forward.[4] Despite the dismissal of similar cases brought by children in other locations, as of this writing, the *Juliana* case remains active.

It's too early to tell whether the litigation strategy of broad-based lawsuits against governments will lead to widespread action, be it to cut emissions or increase resilience. In the cases that have met with success, judges have recognized the existential magnitude of the problem. As the court in the Pakistan case succinctly summarized it, climate change "is a defining challenge of our time."[5] With the physical manifestations of global warming becoming ever more apparent, litigants will undoubtedly continue to push on the courtroom door for relief.

GETTING THE GOVERNMENT TO PAY

A strategy that may bring more immediate results is the pursuit of monetary damages from governments. Although this litigation strategy is still being tested, the mere specter of such judgments can send shivers down the spines of both government agencies and the courts, since liability could be almost limitless. A case stemming from unusually heavy rains in Chicago illustrates a single lawsuit's potential to prompt large-scale change in government behavior.

In April 2013, so much rain pelted the Chicago area that the then-governor Pat Quinn declared a state of emergency. A short time later, the Illinois Farmers Insurance Company (IFIC) filed a class-action lawsuit on behalf of its policyholders. It sued close to two hundred local municipalities and agencies in and around Chicago's Cook County for failing to prepare their stormwater drains and sewers adequately.[6] IFIC alleged that the flood waters in question overwhelmed the systems and pushed water back into people's homes, causing substantial damage. The insurance company claimed that the municipalities should have upgraded their stormwater plans because they knew that "during the past 40 years, climate change in Cook County has caused rains to be of greater volume, greater intensity, and greater duration than pre-1970 rainfall history evidenced, rendering the rainfall frequency tables employed [by the municipalities] inaccurate and obsolete."[7] IFIC demanded that the cities and agencies reimburse it for the payouts it had made to policyholders as a result of the rain damage.

Within a few months, for reasons unrelated to the merits of the case and undoubtedly much to the relief of the municipalities, IFIC changed course and withdrew the case. The complexity and scope of the lawsuit would have made it difficult to prosecute and expensive for all the parties involved, but the lawsuit had sent a clear message. In a press release, the company expressed its hope that the suit would encourage cities and counties to do more to protect against future flooding.[8] Had the suit gone forward successfully, it could have been a landmark case, forcing municipalities to better manage increasing storm water.

The Texas Supreme Court also flirted with making governments pay for their failures to address flooding risk, only to back off when the full implications of doing so became clear. About a year before Hurricane Harvey submerged Houston under more than four feet of water, several hundred homeowners had sued the Harris County Flood Control District, which includes Houston. They claimed that the district had approved development without implementing mitigation measures to address known flood risks, and that flooding had subsequently damaged their homes.[9] Initially, the Texas Supreme Court agreed with the homeowners.

The flood district immediately asked the court to reconsider its ruling and peppered it with more than a dozen friend-of-the-court briefs from organizations representing local governments that opposed the decision. These organizations raised concerns that the court's ruling would open a Pandora's Box of endless liability for governmental entities. One brief described a hypothetical scenario that the Texas Supreme Court later found particularly "disturbing": in a future lawsuit, homeowners could sue the government for damages from a hurricane caused by global warming on the theory that the government had caused the damage by issuing permits allowing the production of fossil fuels and the construction and operation of power plants that burn them. On reconsideration, the Texas Supreme Court concluded that ruling in the homeowners' favor would "vastly and unwisely expand the liability of governmental entities" and, in a rare move, reversed itself.[10] The unfortunate result of this opinion is that Texas counties may continue to approve property development without taking future flood consequences into account.

Similar litigation has been launched in other countries. In 2016, Canadian homeowners sued the government for damages, alleging that the authorities had failed to avert "foreseeable flooding."[11] They claimed that Ontario's Ministry of Natural Resources and Forestry did not take sufficient action to manage higher water levels in several lakes. Historically, the lakes had not been the cause of much flooding, but beginning in 2010, that changed. After heavy flooding occurred in 2016, the homeowners alleged that the Ministry had failed to act when the lakes reached dangerously high levels. The overflowing waters destroyed structures, causing significant losses; the homeowners sought C$900 million in damages. Yet the lead plaintiff dropped the case in 2018 after the Ministry provided documentation that convincingly showed that the lake managers had acted reasonably.

None of these cases has yet delivered the funds the plaintiffs were seeking, but that does not mean they were not successful in one respect. They have put local governments on notice, alerting them to the very real possibility that courts may one day hold governments accountable for inadequately preparing for the new extremes brought on or exacerbated by global warming. This expanding body of climate-related lawsuits may well have significant implications for the choices governments make in a warming world and may lead to greater resilience.[12]

FORCING FOSSIL FUEL GIANTS TO PAY?

Litigation has changed behavior and public policy in the United States before. Many Americans are too young to remember that smoking was once considered benign, perhaps even healthy. People were dying from lung cancer, but the connection between smoking and cancer was not widely understood. It took decades of litigation against tobacco companies to correct the public's perception. Lawsuits against tobacco companies began as early as 1954, when a widow named Eva Cooper sued the tobacco giant R.J. Reynolds claiming it was responsible for her husband's death from lung cancer. She lost the case; for the next forty years, smokers and their families mostly lost such cases against the tobacco companies.

The tide finally began to turn in 1994, when the governor of Florida, Lawton Chiles, signed into law legislation that precluded tobacco companies from defending themselves with the argument that smokers assumed the risk of smoking. The state had concluded that it was time the cigarette manufacturers paid for smoking-related illnesses rather than the state of Florida. Other states followed, and smokers soon began to win in court. These legal efforts culminated with forty-six state attorneys general uniting in litigation to recover the rising public health costs from smoking. Eventually, the tobacco companies and the states settled for a mind-boggling $246 billion, the largest settlement in US history.

It took close to a half-century of legal wrangling for a smoker to win a significant verdict against a tobacco company. The smokers first had to overcome numerous legal hurdles, including the need to trace specific damages to specific manufacturers and to prove that a particular tobacco company "caused" the injury. Smokers also had to contend with conflicting scientific reports and the notion that they "accepted the risk" of smoking. But they were finally able to surmount these hurdles. Public understanding of the dangers of smoking increased dramatically, as did government efforts to protect public health. In subsequent years, rates of smoking decreased markedly, with visible gains in the health of the population.

Those seeking damages for climate change impacts from fossil fuel companies have sought to follow the tobacco playbook. In 2017 and 2018, several subnational governments in the United States sued the fossil fuel companies for billions of dollars in damages, just as little Kivalina had in 2008. When explaining why New York City sued, Mayor Bill de Blasio made explicit reference to the tobacco litigation: "We wanna follow that same path, help to end climate change impacts by changing [fossil fuel companies'] behavior and get money back for the people of [New York]."[13] Similarly, the former governor of California Arnold Schwarzenegger observed that suing the fossil fuel companies "is no different from the smoking issue," and declared his intention to go after them "like an Alabama tick."[14]

At the very least, he said, the lawsuits would raise awareness and encourage people to look for alternative fuels.

Fossil fuel companies have fought back hard, framing the battle as one that threatens the very underpinnings of democratic government. After the judge dismissed the case brought by New York City, Theodore J. Boutros, one of the defense lawyers, made his relief plain. He praised the judge for getting "it exactly right . . . Trying to resolve a complex, global policy issue like climate change . . . would intrude on the powers of Congress and the executive branch to address these issues as part of the democratic process."[15] Those suing fossil fuel companies will need to play a long game. Much is at stake, including the possibility that the first verdict in favor of a plaintiff will "set off a cascade of lawsuits" costing the defendants billions of dollars.[16]

Although the litigation has not succeeded in American courts as of this writing, at least one court has found fossil-fuel emitters liable for damages caused. Germany's largest power company, RWE, found itself on the losing end of a lawsuit filed in German courts by Peruvian farmer and mountain guide Saul Luciano Lliuya.[17] Though the power company did not even operate in Peru, Lliuya sought €17,000 (roughly $20,000) worth of flood protection to pay for what he estimated was the utility's share of responsibility for the climate impacts where he lived. He claimed that the utility's carbon emissions had contributed one half of one percent of global emissions. The German company should therefore pay one half of one percent of the cost of protecting against the glacial melt that threatened to flood his community. A lower court had found that the case was too weak and blocked it from going forward. But in 2017, the Higher Regional Court in Hamm disagreed and allowed the case to proceed. As of this writing, the case is still active.

In an ironic twist of fate, fossil fuel companies may also find themselves defending climate lawsuits unrelated to carbon emissions. These corporations are potentially on the hook for failing to protect their own facilities adequately against sea-level rise and extreme weather exacerbated by warming temperatures. Shell Oil, for example, found itself in this situation with regard to a bulk storage and fuel terminal it operates in Providence, Rhode Island.[18] In Boston, Massachusetts, Exxon also faced a lawsuit for failing to take measures to protect its distribution and bulk storage terminal from climate impacts.[19]

LIABILITY AT YOUR DOORSTEP

In the coming years, as the severity and frequency of climate change-related damages grow, a new group of people will likely find itself on the receiving end of litigation alongside fossil fuel companies and governments. That group includes corporate directors and officers, architects, engineers, and others who have a duty to consider foreseeable harm and to manage the risk.[20] At issue will be whether they acted reasonably in the face of risks they knew or should have known about. If harm from warming temperatures is foreseeable, and professionals ignore the risk because they do not believe in climate change, they may find themselves sitting in the defendant's chair in a courtroom. At the greatest legal risk will be professionals who rely only on historical experience to make decisions even when credible authorities—such as the authors of the US Fourth National Climate Assessment, a congressionally-mandated analysis issued in 2018 identifying climate risks to the nation—assert that it is no longer valid to assume that future climate conditions will resemble the past.

In such cases, judges and juries may begin to find that architects, planners, and engineers acted unreasonably when they made consequential decisions based on outdated information. For example, using building codes and standards that do not reflect the current and future risk of climate-exacerbated extreme weather may be viewed as falling below the standard of care required by professionals. Similarly, judges and juries might begin to find that private owners and operators of airports, toll roads, and utilities should have done more to protect infrastructure and its users given known risks from climate change. Over time, as the impacts of climate change become more obvious, judges and jurors may feel less sympathetic toward people in positions of responsibility who put lives or livelihoods in harm's way when they should have known better. The long history of tobacco litigation illustrates how the law adjusts to new information and science over time, and how that adjustment can lead to changes in the duty of care owed by professionals.

The looming threat of liability will command ever-more attention from corporate officers, directors, and their insurance carriers as well. For example, corporate directors in Australia are already feeling the heat. A 2016 legal analysis concluded that directors would be well advised at least to consider the possible effects of "climate change risks" to their businesses.[21] If sued, directors and officers facing lawsuits will likely approach their insurers seeking compensation and reimbursement for lawyers' fees. If there are gaps in insurance coverage, corporate officers and directors may even find themselves personally on the hook for damages.

The science of climate attribution could also reshape our understanding of climate change liability. "Climate attribution" refers to scientists' growing capacity to quantify the extent to which climate change magnifies the severity or likelihood of events such as extended droughts, extreme rainfall, and heatwaves. Scientists can now determine with increasing confidence whether a particular event was more likely because of climate change, and even whether the event would have occurred at all without climate change. For instance, the American Meteorological Society issues annual reports devoted to explaining extreme events from a climate perspective. In its 2016 report, the society included, for the first time, scientific findings showing that three events—extremely high temperatures in Asia, global record heat, and a marine heatwave off the coast of Alaska—would not have occurred absent climate change.[22] Over time, improvements in attribution science will find their way into courtrooms, and when they do, professionals will have to think twice about their decisions in a world impacted by climate change. The growing risk that judges and juries will find these professionals liable will undoubtedly push corporate leaders and professionals to consider those impacts more fully in their decision-making, leading to more investment in resilience.

In addition, liability may extend beyond paying damages—it could include criminal prosecution. French courts provided a glimpse of this risk in the aftermath of Cyclone Xynthia, which slammed into the French Atlantic coast in the dead of night in 2010. The storm killed almost fifty people and caused more than €1 billion in damage. Twenty-nine of the fatalities occurred in just two small villages, L'Aiguillon-sur-Mer and La Faute-sur-Mer. As the storm hit the villages, the waters broke through a seawall, sweeping into a low-lying depression which contained many single-story homes. The next day, France's president, Nicolas Sarkozy, visited the site of the disaster and launched an investigation. In his words, "We have to find out how families in France in the twenty-first century can be surprised in their sleep and drown in their own houses."[23]

The government eventually filed criminal charges for manslaughter against the four-term mayor of La Faute-sur-Mer and his deputy. The court sentenced the mayor to four years in prison and his deputy to two years, finding that they "deliberately hid" the known flood risks to take advantage of the property development "cash-cow."[24] The court was not persuaded by the mayor's argument that he and the town council lacked the technical resources to evaluate storm risks.

Criminal investigators also took a dim view of the role a Florida nursing home played in the death of twelve patients. In 2017, Hurricane Irma caused the facility to lose power, leaving the elderly residents in deadly heat. The police opened a homicide investigation. If more jurisdictions choose to pursue criminal charges after future disasters, they could spur government officials and others to do better at preparing for foreseeable harm caused by climate-exacerbated extreme events.

SMALL CASES, BIG IMPLICATIONS

Though the evolution of climate litigation remains unclear, it's safe to say that the number of cases will grow, posing a set of legal issues that courts have never confronted before. What should the responsibility of the judiciary be as increasing temperatures and their consequences upend our understanding of responsibility, liability, and justice? What sorts of precedents could be set by seemingly minor lawsuits? As the judicial system struggles with these issues, lawsuits have already begun to illuminate how a single piece of litigation has the potential to drive or hold back resilience.

Take, for example, the question of what a local government should do when sea-level rise repeatedly damages a section of road within its jurisdiction. Florida's St. Johns County faced such a dilemma in 2004, when a hurricane washed away a stretch of road.

The County Commission faced a difficult choice: repair the road it had already been defending from rising sea levels at a cost of one million dollars, or let it slip into the sea.[25] The county decided it would no longer spend the money to maintain the road. But the decision had very real consequences for Summer Haven, a small community that could only be reached via that road. Summer Haven residents sued, and the court sided with them, finding that once the county had undertaken to maintain the road, it had a duty to do so reasonably.[26] The court did not have a chance to decide what constituted "reasonable" maintenance before the case was settled out of court. Had the court decided the issue, the answer could have had profound implications for the many government-maintained roads that are at risk of erosion from sea-level rise.

Consider also what happened in the Harvey Cedars, a community of just a few hundred people that sits on a barrier island off the coast of New Jersey. In 2010, Harvey Cedars decided it wanted to take advantage of a $25 million flood-protection project sponsored by the government. As part of the project, the local government built a twenty-two foot dune on the property of local residents Harvey and Phyllis Karan (see Figure 2). The Karans and the local authorities could not agree on the amount of fair compensation owed to them for the use of their property and for the loss of their oceanfront view. Harvey Karan complained that in all the years he had owned the property, he had never seen "a lick of water" reach the living quarters.[27] The borough and the homeowners ended up in court.

Following existing law, the trial court excluded evidence that the dunes would greatly reduce the risk of damage to the Karans' home for the next thirty years, and the jury awarded the couple $375,000 for the loss of their oceanfront view and use of their lot. On appeal, the New Jersey Supreme Court reversed the decision, concluding

Figure 2 Dune protection in front of the Karans' home in Harvey Cedars, New Jersey. Source: Wayne Parry/Associated Press.

that relying on the existing law had allowed homeowners to be paid unfairly considering that their house would also benefit from the dune protection. The New Jersey Supreme Court fashioned new law allowing for compensation that is "just to both the landowner and the public."[28] In 2013, the Karans settled for one dollar in damages.[29]

What happens when homeowners want to protect themselves from climate impacts, but local zoning ordinances stand in the way? When Hurricane Sandy swept through Milford, Connecticut, in 2012, it caused so much damage to a hundred-year-old house that the homeowners had to demolish it. When the homeowners went to rebuild, they wanted the new structure elevated to protect it against future flooding. The City of Milford, however, had a zoning regulation that restricted the height of homes for aesthetic reasons and refused to grant them an exception to the local ordinance. The case ended up in court, and the judge made the court's displeasure with the city clear, noting that it is "an abiding principle of jurisprudence that common sense does not take flight when one enters a courtroom."[30] The judge found that aesthetic height regulations did not outweigh the importance of elevation for public-safety reasons and deemed the city's policy "contrary to law and logic."

Courts will also be asked to apply existing law to new circumstances resulting from warming average temperatures. For example, a federal court in Texas found that the lack of air conditioning in the state penitentiary constituted cruel and unusual punishment, which is prohibited by the US Constitution. Prisoners described the conditions in the prison as equivalent to "walking out to your car in the middle of the summertime" and "getting into a hot box in the sun."[31] Importantly, the court ruled that even though the prison had "no way of knowing when a heat wave will occur," it needed to take action because "it is clear that one will come."[32] In the end, the inmates got air-conditioning. With average temperatures continuing to rise across the United States, these concerns will affect more and more correctional facilities that are ill prepared for the new heat extremes.

Virginia Beach, Virginia, a city adjacent to Norfolk that suffers from sea level-rise-related flooding, provides another example. The Virginia Beach City Council unanimously denied a building permit for over two dozen new homes in an area prone to flooding on the grounds that the project developer had not addressed the flood risk adequately. The developer sued, claiming that the city had unfairly increased stormwater-management requirements in the midst of his permitting process. The council countered that it had acted reasonably to protect against future flooding. If the development went forward without proper stormwater protection, the cost of remediating the flood could fall on the city, not the developer. The Virginia courts are still considering the matter as of this writing.

As these cases reveal, litigation will drive—and sometimes impede—on-the-ground efforts to build resilience. Whether or not climate change litigation results in a seismic shift in the law, it is certain that such litigation will occupy ever-greater space on the already bulging dockets of trial courts in the United States and other countries. A 2017 report from the United Nations Environment Programme found that before 2014, climate change-related cases had been decided in only twelve countries. By 2017, that number had increased to twenty-four countries. The report also concluded that climate change litigation appeared to be growing in both ambition and effectiveness, as more litigants were holding governments accountable for their actions and inactions related to climate change impacts.[33] The promise of future litigation surely lies behind the advice of a respected resilience expert, who told a large Washington audience in 2017, "If I were to advise my children about what career to go into related to climate change, I would say: 'Be a lawyer.'"

DODGING THE BIGGEST ISSUE OF ALL?

Some judges feel the climate change challenge is just too big for the courts to take on. In dismissing the damages lawsuit brought in northern California by the cities of San Francisco and Oakland against fossil fuel companies, federal Judge William Alsup concluded: "The problem deserves a solution on a more vast scale than can be supplied by a district judge or jury."[34] In his opinion, the problem demands "the expertise of our environmental agencies, our diplomats, our Executive, and at least the Senate."[35] In New York City, a federal judge handling Mayor de Blasio's suit against the fossil fuel giants reached a similar conclusion. Judge John Keenan found that "the serious problems caused [by climate change] are not for the judiciary to ameliorate" and that they "must be addressed by the two other branches of government."[36]

But others take a different view. Some judges and commentators argue that in view of politicians' failure to act, the judiciary should respond and that the courts should not dodge this responsibility. The trial judge in *Juliana v. United States* concluded that the case raised a constitutional violation, an issue that lay "squarely within the purview of the judiciary."[37] She criticized the federal courts for being overly cautious and deferential when it came to environmental damage, an approach that had left the world to suffer. Judge Alfred Goodwin of the Ninth Circuit Court of Appeals provided an even more damning assessment, accusing the legal system of a "wholesale failure . . . to protect humanity from the collapse of finite natural resources by uncontrolled pursuit of short-term profits."[38] He urged courts to re-examine whether they had created too many barriers to achieving justice in environmental cases. History will judge whether US courts missed a key opportunity to help avert humanity's impending collision with a future drastically changed by global warming. Are judges simply doing their duty, or are they dodging the biggest issue confronting the planet, or both?

Even if they fail to resolve the essential challenge of cutting greenhouse-gas emissions, the courts will surely find themselves enmeshed in litigation for years over who pays for the damage. They may help homeowners understand that flood protection is in their interest, and they may force local governments to amend their codes to permit common-sense resilience measures. In courtroom after courtroom, judges will reach decisions that can contribute to resilience on a large scale.

PRESCRIPTIONS AND PROVOCATIONS

- Law and judicial associations should provide training to lawyers and judges on climate science, attribution, emissions reduction, and climate resilience measures. The American Law Institute should examine whether their authoritative treatises appropriately address the emerging legal issues presented by climate change.

- The federal government should establish a national commission to recommend possible amendments to liability laws to help drive more resilience.

- Criminal prosecution offices should increase the publicity for successful prosecutions for failure to prepare adequately for climate impacts to deter others from recklessly endangering the public.

[3]

MAKE MARKETS WORK
FOR RESILIENCE

As the global financial system descended into crisis in 2008, the Canadian government appointed a forty-three-year-old finance official named Mark Carney to head its central bank, making him the youngest central bank governor of any major economy. Carney's first mission was to enable his country to navigate some of the most turbulent waters in modern economic history. With a series of bold and well-timed policy moves, the governor helped Canada to emerge from the crisis relatively unscathed. Carney's performance so impressed monetary authorities abroad that five years later, he was appointed governor of the central bank of a different country—the United Kingdom.

In some ways, Carney was an unusual pick to head the Bank of England. Raised in Canada, educated in the United States and Britain, married to a Briton, and holding Canadian, British, and Irish citizenships, the new governor had a more cosmopolitan background than his predecessors, going back to the central bank's founding in 1694. With his penchant for creative thinking, some of Carney's ideas were bound to be outside the box.

In September 2015, he gave an unusual speech in London to a group of tuxedoed insurance executives. Carney warned the assembled crowd about a risk that central bankers had largely ignored but that kept him up at night. "Shifts in our climate bring potentially profound implications for insurers, financial stability, and the economy," he said.[1] Decision-makers in business and government think and operate on time horizons that are too short to capture some of the worst impacts of climate change. This makes it hard to get a handle on the problem. Carney called this "the tragedy of the horizon", and escaping it, one of the central challenges of our time.[2]

Carney's proposed solution was seemingly simple: disclose climate risk. Companies should disclose to the public what they know about the risks they face from climate change and how they plan to deal with them. These include the risk that policy and technology will shift quickly as the world tries to cut carbon emissions, upending a company's business model. But they also include the risks that are the subject of this book—namely, the physical impacts from climate change that can disrupt businesses and entire economies.

As more and more information about climate risk becomes available to the public, Carney argued, the prices of all sorts of assets traded in markets—stocks, bonds, property—will shift to reflect the true risk of climate change. And as prices shift, so will behavior. In a world in which climate risk is plainly transparent to all, everyone will be pressured to manage the risk and protect the value of their assets. Companies will abandon business models that generate high carbon emissions. Developers will stop building in flood zones. Investors will put their money only in companies that demonstrate they are serious about managing the risks posed by climate change. Market forces will work in favor of climate action and resilience on a large scale. That's the theory, at least.

This chapter looks at how disclosure can make markets work for resilience and examines the challenges and dilemmas that this entails. We look at four markets in which we might expect climate risk disclosure to most readily cause prices to change. The markets are equities (company stocks), debt (bonds issued by companies and governments), property (real estate), and insurance. The insights learned from these markets can illustrate how each could drive resilience on a large scale.

LEARNING TO LOVE DISCLOSURE (THE EQUITY MARKET)

Before the US stock market crashed in 1929, it resembled the Wild West. Wall Street operated like a rigged casino, a largely unregulated space with rampant fraud and abuse. Lack of disclosure was a big part of the problem. A former regulator described it this way: "Before the crash, stock prices often had little to do with the fundamentals, because most of the fundamentals were never disclosed . . . Investors were sold securities without the benefit of a prospectus or offering circular; without ever seeing a balance sheet; without knowing the first thing about a company beyond its name and share price."[3]

To build trust in the stock market, the US Congress passed the 1933 Truth in Securities Act and created the Securities and Exchange Commission. These measures changed the game, laying the foundation of a stock market based on disclosure—the notion that those selling stocks to the public must disclose "material" information, or information that is likely to change the perceived value of a security when it is known to the public.[4] "Materiality" has served as a guiding light of American financial disclosure ever since, helping to cement widespread trust in the US capital market. Today, debates are raging about whether climate change and responses to it pose material risks to companies. For example, as the world moves to cut carbon emissions, companies whose business depends on high-carbon-emitting activities may soon find that their business models no longer work. Most importantly for our purposes, companies may find their operations, supply chains, and profits undermined by the effects of climate change.

That the extreme weather, sea-level rise, and other phenomena exacerbated by climate change pose material risks to companies is clear. Consider what happened to major corporations with operations in Thailand in 2011. That October, early in the monsoon season, a steady rain fell over northern Thailand. These rains grew to be one-and-a-half times more intense than normal. The Chao Phraya River system overflowed, and water reservoirs designed to contain the flooding breached their limits. Soon the waters reached seven industrial parks surrounding the capital city of Bangkok, which contained hundreds of factories belonging to eight hundred different companies, about half of them Japanese. Some of those factories housed producers of key auto components, such as power integrated circuits, audio and navigation systems, and transistors and condensers. The floods paralyzed production. Faced with a shortage of critical parts, the companies' supply chains, which stretched from Thailand across Southeast Asia and beyond, seized up. It would take two months to pump the water out of all the flooded facilities. Some never reopened. Between them, Japan's big three automakers—Toyota, Nissan, and Honda—experienced production losses of 420,000 cars, which cost them about $2.8 billion in operating profits.[5]

The electronics sector was also hit hard—especially the makers of hard disk drives. At the time, Thailand produced over 40 percent of the world's hard disk drives, essential components in every personal computer and handheld device. Western Digital Corporation alone, which produced one-third of the world's hard disk drives in 2011, lost half its shipments in the flood. Factories owned by Toshiba, Samsung, and Seagate Technology cut production as well. As production dried up, a shockwave spread across the world's computer market, and the prices of hard disk drives doubled and, in some cases, tripled. Computer manufacturers soon felt the cascading effect on their supply chains, prices, and profits.

Other industries are also at risk from climate change. The global shipping industry is vulnerable, as sea-level rise can render ports and warehouses inoperable.[6] Software and data-management companies run the risk that superstorms will flood key data centers. Airlines face the prospect of having to cancel flights when high temperatures prevent their planes from taking off, and airports located at sea level could suffer from flooded runways. Banks holding lots of mortgages clustered in the same geographic area could suffer if many homeowners stopped making mortgage payments at the same time, because perennial flooding has made their homes valueless. The power-generation sector can get hit by water shortages; when there is not enough water to cool the turbines, power plants must shut down.[7] Wildfires that damage electrical infrastructure can disrupt electricity transmission. Yet companies are still not disclosing these risks systematically and consistently to shareholders and the public at large.

To address this issue, a panel of business leaders came together in 2015 with the blessing of global financial regulators. The panel, known as the Task Force on Climate-Related Financial Disclosures (TCFD), offered sensible and seemingly simple advice.[8] Companies should identify and disclose their climate risks. They should explain the impact such risks could have on their strategic direction and financial health. They should engage in scenario planning and disclose how the business might fare under different climate scenarios.

And they should explain what they intend to do about the risks (and climate-related opportunities) they have identified, assuming they find these to be material.

Some regulators and hundreds of business leaders applauded and endorsed the TCFD recommendations, but implementation ran into two major obstacles. First, assessing and describing physical climate risk is a complex undertaking, especially if companies are to report in a comparable way. Those doing the disclosing need to agree on what to report and how to report it. How should companies measure and report the risk of water stress, wildfires, or floods? What are the most appropriate indicators and time horizons to use in measuring the risk? What assumptions and scenarios should companies adopt when reporting? If they can't agree and every company discloses using different metrics and assumptions, the market will not know what to do with the cacophony of information and will likely ignore it. Prices won't shift, and behavior won't change.

Fear is the second obstacle. As long as climate-risk disclosure remains voluntary, companies will worry about a "first-mover disadvantage"—that is, that the market will punish those who disclose their climate risks first, putting them at a disadvantage relative to competitors who choose to stay silent. Some corporate leaders feel that it's better to stay mum until regulators make climate-risk disclosure mandatory. At the same time, they are working quietly to understand climate change risk, so that their companies can manage it more effectively and be ready if and when mandatory disclosure finally arrives. One banker at a private gathering in New York City in 2016 put it best: "If I were a corporation, I would be denying climate risk while preparing for it in secret."

A related issue is legal liability. Corporate lawyers worry that if a company is found to have known about risks from climate change impacts but failed to report them, it will face liability for suppressing the information. Climate disclosure advocate Stan Dupré has talked to dozens of companies on both sides of the Atlantic about climate risk and has a good sense of the problem. Dupré typically starts exploring climate risk with a firm's executives, and eventually the question of whether the company should disclose climate risk as part of its legally required financial disclosures arises. "The conversation starts with the legal team," says Dupré, "and basically the answer we get usually is, 'OK, if we start disclosing this risk and our competitors don't, we will lose, so we won't do it . . . Not only it won't be reported [sic], but we should stop investigating this risk, otherwise we will lie to our investors."[9] This is hardly what Carney had in mind.

One solution is mandatory disclosure, which would force everyone to disclose at the same time against the same standards. This would eliminate the fear of first-mover disadvantage, at least within a given country, and it would get the market and regulators accustomed to digesting this information and incorporating it into their decision-making. Europe is leading in this space. France has already mandated the disclosure of certain climate-related information. In 2018, the European Commission indicated that it will provide more guidance to companies on how to disclose climate-related information in line with the TCFD recommendations.[10] It launched a process that could mean that, before long, all major companies operating in Europe will have to disclose their climate risks. Making disclosure mandatory in other key economies, especially in the United States, Japan, and China, will be important to ensure that companies in places with looser disclosure requirements don't benefit unfairly at the expense of those in countries with tougher disclosure rules.

PAYING BACK IN A WARMING WORLD
(THE BOND MARKET)

One spring day in 2015, the White House received a call from an anxious official in the city of Norfolk, Virginia. As we have seen, Norfolk is on the front lines when it comes to sea-level rise. The credit-rating agency Moody's, which rates Norfolk's bonds, had sent a detailed questionnaire to city hall. The message, reading between the lines, was clear. How, Moody's wanted to know, does the city plan to address sea-level rise and recurrent flooding in the face of climate change and still pay back its debts? Apparently, an article in the *Washington Post* had piqued Moody's interest. The article reported that a plan commissioned by the city projected that protecting Norfolk from sea-level rise would cost $1 billion—almost the size of the city's entire annual expenditure budget.[11] Norfolk was not alone. Portsmouth, Hampton, and Virginia Beach, coastal cities in Virginia adjacent to Norfolk, received the same letter. The Norfolk official worried that Moody's, concerned about the city's creditworthiness, might downgrade the city's bonds.

The official was right to be worried. Credit-rating agencies play a powerful role in the bond markets. Companies and governments issue bonds in order to borrow money from investors. The investors who buy the bonds expect bond issuers to repay the principal with interest over a certain period of time. Credit-rating agencies like Moody's rate the likelihood that an issuer will be able to pay back its investors on time. If Moody's downgraded Norfolk's bonds, the market would likely see the city as a riskier bet and demand a higher interest rate to compensate for the higher risk. Norfolk would then find it more expensive to raise the money to pay for all kinds of things, including investments in resilience.

It's not just at-risk cities like Norfolk that worry about what climate change impacts could do to their ratings and borrowing costs. Many poor countries and small island nations are concerned that their exposure to climate risk has already pushed those costs higher, making it harder for them to finance much-needed resilience investments. A team of London-based researchers reckoned in 2018 that climate vulnerability has in fact pushed up borrowing costs for many developing countries. The researchers estimated that these countries pay, on average, about 1.17 percent more in interest based only on their climate vulnerability, which amounts to $40 billion in additional interest payments over the past decade on government-issued bonds alone.[12]

Yet despite the concern of officials in Norfolk and some developing countries, credit-rating agencies have been slow to capture climate risk. "Climate downgrades," or situations in which a climate impact or climate risk has unambiguously led to a credit downgrade, are hard to spot. Sometimes, issuers continue to enjoy stable ratings for a long time in spite of escalating climate risk. Norfolk's bonds remained stable long after the city official's anxious call to the White House. And Miami, for example, continues to benefit from relatively low borrowing costs, which seems jarring considering the existential climate risks the city faces over a longer time horizon.

One reason why climate risk doesn't show up clearly in credit ratings is that the ratings take into account many factors in addition to climate impacts. Another reason is that the time horizon of the ratings is too short—typically, five years at most—to capture the more severe impacts of climate change, which take longer to emerge. Sometimes, when a disaster hits a relatively well-off community, the rating agencies even see it as a positive development. "Some of these disasters—it's going to sound callous and terrible—but they're not credit-negative," one senior credit-rating agency executive told the press. The communities "rebuild, and the new facilities are of higher quality and higher value than the old ones."[13] Communities might be better off after a disaster, sitting on top of more valuable assets that enhance their creditworthiness, at least until the next disaster strikes.

Precisely because credit ratings sometimes fail to capture climate risk, some see an opportunity to make money. Hedge funds and other investors are using computer models (some of which we will discuss in chapter 5) to estimate which US municipalities will face relatively low climate risk in the future. Then they buy the bonds issued by those local governments. Because the market does not yet distinguish very well between the bonds of high-risk and low-risk municipalities, the hedge funds can gain an advantage and buy low-risk bonds relatively cheaply. Eventually, when mandatory disclosure or the forces of nature reveal to everyone the places that are more exposed to climate risk, the bonds of the low-risk municipalities will rise in value, and those of the risky ones will sink. Hedge funds could then sell their lower-risk bonds at a premium and pocket the difference. At the same time, once the bond market wakes up to the reality of climate risk, the most vulnerable communities will have to pay investors more to borrow money.

How should the Norfolks of the world react? They should use this time, before bond prices fully reflect climate risk, to develop a strong resilience plan and start implementing it right away. If companies, municipalities, and whole countries can demonstrate to the market that they have a plan and are putting measures in place to reduce the risk, it should help differentiate them from peers who may be sticking their heads in the sand. Taking action now may protect their credit ratings and borrowing costs, helping them continue to make investments in resilience.

PROPERTY AT FLOODED-BASEMENT PRICES (THE REAL-ESTATE MARKET)

The United States, along with Canada, Australia, France, and the United Kingdom, has one of the most transparent property markets in the world.[14] Most US states require public disclosure of data on property prices, ownership, and sales. Technology has put that huge volume of data into the public's handheld devices at no cost. Zillow, one of the largest real-estate search engines in the US, has 110 million properties in its database alone and nearly 200 million monthly visitors.[15] Trulia, a search engine acquired by Zillow, developed a function that allows would-be homebuyers to layer property maps on top of maps showing historical data on hurricanes, wildfires, tornados, and floods. In theory, the real-estate-buying public should be more aware than ever about growing climate risks. In light of increased information, do we see climate resilience and vulnerability reflected in real estate prices yet?

In the United States, the answer is yes. Consider that property prices in the areas affected by Superstorm Sandy dropped between 6 percent and 16 percent, even for properties that were not directly damaged by the storm. Some worried that the price signal would fade along with memories of the storm. But four years later, economists could still pick up the price discount.[16] A similar story emerges from nationwide data. Researchers looked at data from nearly a half million US property transactions and compared it with projected sea-level-rise data.[17] Coastal properties exposed to projected sea-level rise sold at a discount of around 6 percent to 7.5 percent compared to similar properties that were not threatened by sea-level rise. The researchers also found that the discount was larger in places where public awareness of sea-level rise was higher. Similar results have emerged in the Netherlands, where property prices are 1 percent lower, on average, in places at risk of flooding, and the discount gets larger in neighborhoods with higher flood-level projections.[18]

Not only are buyers paying less for properties in flood zones, but evidence is beginning to mount that they are paying *more* for homes in locations perceived to be resilient. In Florida's Miami-Dade County, researchers have found that the higher the elevation of single-family properties, the more rapidly their prices have appreciated. Also, the prices of properties located at sea level, and thus more likely to experience chronic flooding, have not kept up with the price appreciation of homes at higher elevations.[19] As property buyers act on their perceptions of climate risk, they appear to be moving prices, and those prices, in turn, are signaling to others that they should think about climate-safer places when shopping for a home.

Some players in the real-estate market want to promote climate-risk awareness and are pressing for greater availability of information about risk. Zillow, for instance, is making its massive database of property transactions available to researchers studying climate risk.[20] Also, some jurisdictions have made it mandatory for property owners to disclose flood risk when selling a property. In Miami-Dade County, local laws require that owners disclose to buyers in writing—in at least ten-point boldface type—if the property falls within a special flood-hazard area.[21]

Despite growing information about potential climate impacts to property, real-estate prices still don't reflect the true underlying risk in many parts of the United States. Prices have not yet stopped developers from building luxury towers on potentially vulnerable coasts (Figure 3) and residential complexes in areas at extreme risk of wildfire, nor are they keeping customers from buying them. In post-Sandy New York City, new construction continues to go up in the areas that were the worst hit by the storm. As of 2018, there

Figure 3 Rendering of a new luxury tower in Miami, Florida.

were over twelve thousand new apartments planned or under construction in the city's worst flood zones; roughly one in eight new
apartments in the city will sit in a high-risk flood zone.[22]

As we saw with the equity market, fear moves some to try to delay the
day of reckoning. Property owners, municipal governments, and realtors
all have something to lose from too much climate-risk transparency and
from the wrenching shift in prices, profits, and taxes that will result from it.
After Katrina, New Orleans, for example, lobbied the Federal Emergency
Management Agency (FEMA) for seven years to change its flood maps
and exclude many city properties from the high-risk area. And after
FEMA revised New York City's Flood Insurance Rate Map and added

tens of thousands of properties to the "highest risk" category, city leaders lobbied hard to keep those properties out of that category, eventually with success. Nevertheless, it seems that property markets are waking up, faster than equity and bond markets, to the reality of climate risk.

THE COST OF PROTECTION (THE INSURANCE MARKET)

The fourth market we examine is the market for insurance, especially property insurance. Everyone who owns property should want to buy protection in case their property suffers damage from water or fire. Insurance companies are happy to sell that protection for a price. That price reflects the likelihood that the property will suffer damage, to the best of the insurance company's calculations. This is why insurance companies employ thousands of actuaries, whose job is to calculate probabilities and prices. Insurance buyers are of course free to shop around for the cheapest insurance rate. Homeowners with more resilient properties, or properties in areas with lower risk of climate-related extreme events, should pay less for their insurance than those who own riskier homes. In theory, the insurance market should be incentivizing resilience on a large scale.

Yet this does not always happen in the United States. The history of flood insurance helps to explain why. Flooding is the most common and costly threat to property in the United States. In the 1960s, successive flooding events caused many private insurance companies to back away from selling flood insurance in certain parts of the country because they viewed it as too risky. In response, Congress created the National Flood Insurance Program (NFIP) in 1968 to provide affordable insurance to communities facing significant risk of flooding. The NFIP is now the primary flood insurer in the United States. It provides insurance coverage for properties nationwide, including many that are highly vulnerable to flooding. Private insurance companies do not offer coverage for these vulnerable properties in part because few property owners can afford the premiums. Instead, the NFIP sells insurance to some property owners at discounted prices, ultimately transferring the risk of losses to the federal government.

The NFIP wasn't supposed to work this way. Originally, the government intended for the NFIP premiums to reflect homeowners' true flood risk. But to keep premiums low for certain groups, Congress overrode this principle and required the NFIP to charge different categories of homeowners less than the true risk would demand. As a result, for about 20 percent of NFIP policyholders, typically the ones in highest flood-risk areas, the program charges premiums that do not reflect the true risk. When those properties suffer damage, the federal government absorbs a good part of the cost, often repeatedly. As climate change exacerbates flooding from sea-level rise and extreme precipitation, this burden will likely balloon. At the time of this writing, the NFIP is over $20 billion in the red, even after Congress forgave $16 billion of its debt in 2017.[23]

Aware of the growing shortfalls, Congress has tinkered with the program over the years and, in 2012, it tried to put the NFIP on sounder financial footing. Congress passed legislation, with overwhelming bipartisan support, that eliminated subsidies and allowed the program's insurance premiums to rise gradually over five years until they fully reflected the true risk to the insured properties. But once the bill was law, it became clear that the measure would trigger large premium increases for certain homeowners. Stories about these homeowners began to appear in the media. The *New York Times*, for example, wrote about a firefighter who expected his premium to jump from under $500 a year to as high as $15,000 a year.[24]

Public outrage erupted, and the political pressure on legislators became too much. Maxine Waters, one of the key sponsors of the reform, conceded to the press that, "[n]either Democrats nor Republicans envisioned [that the legislation] would reap the kind of harm and heartache that may result from the law going into effect."[25] In a spectacular reversal just two years after the original law passed, Congress approved, with equally overwhelming bipartisan support, new legislation that undid much of the reform. The new legislation repealed some provisions of the original law, grandfathered some of the NFIP subsidies, and ordered an 18 percent cap on annual premium rate increases.

A key lesson from this debacle is that once governments subsidize insurance premiums to shield property owners from the true cost of their flood risk, it becomes very difficult to shift back to a free market in which the price of insurance fully reflects the risk. If there is an attempt to do this, it cannot be done abruptly. The British government applied this lesson in 2016, when it introduced Flood Re, a system that subsidizes the flood-insurance premiums of property owners in flood-prone areas. Crucially, the system is set to expire in twenty-five years, at which point the subsidy will end. This timeline gives homeowners a transition period to put in place resilience measures and prepare for the day when premiums fully reflect the risk. At the same time, the system does not offer insurance coverage for new properties, so if developers want to build in risky areas, they bear the full risk of doing so. The United States and other countries should consider adopting a similar approach.

What about risks the private market currently insures against that will be amplified by climate change? Take the risk of wildfires, for instance. Of the millions of California homes in the wildland-urban interface, one million are rated at high or very high risk of wildfire. Reeling from several years of epic wildfire losses, private insurers have increased fire-insurance premiums in California, and some have considered pulling out of the market altogether. When private insurance dries up, homeowners can turn to California's Fair Access to Insurance Plan (FAIR Plan), a state-mandated insurance pool created after brushfires in the 1960s made insurance hard to obtain. The plan is run by private insurers for homeowners who cannot get insurance elsewhere. "The FAIR Plan is the canary in the coalmine," said California Insurance Commissioner Dave Jones in 2018. "If you see their numbers begin to go up dramatically, that tells us people are not able to find private insurance, and it tells us the private insurers are really pulling back."[26] As the number of homeowners covered by the FAIR Plan grows, it's not clear how long private insurers will be willing to continue to participate without the government stepping in to shoulder some of the risk. Then the prospect of another costly NFIP-like arrangement, this time for wildfire risk, will grow more and more probable.

One way to make insurance markets work for resilience involves strengthening standards and practices. Several nonprofits in the United States are developing building-safety standards that can demonstrably reduce the risk of damage to a property. If a homeowner meets the standard by undertking retrofits and upgrades, and an independent authority can verify that the property does in fact meet the standard, then insurers may be willing to lower their insurance premiums. The property owner can use some of the insurance savings to pay back loans used to pay for the upgrades.

For example, a corporation that operates in several states called MyStrongHome encourages homeowners to upgrade their properties to meet extreme-weather resilience standards. A nonprofit organization called the Insurance Institute for Business and Home Safety (IBHS) develops safety standards based on research it conducts in a unique facility in South Carolina, where its experts test different protective technologies. In its hangar-sized laboratory, IBHS can subject one and two-story model homes to simulated hurricane-force winds, extreme rainfall, and even ember showers. Once MyStrongHome certifies that a structure meets IBHS standards, participating insurance companies offer the owner a discounted premium on his or her policy. A program in Colorado called Wildfire Partners has done something similar. It has developed standards and best practices for protecting buildings from wildfires, and it certifies homes that comply with those standards and practices.[27]

These efforts, however, have struggled to reach mass scale. Wildfire Partners certified less than a thousand homes in Colorado between 2014 and 2019. And in Alabama, the state with the most IBHS-compliant homes, only seven thousand houses were certified between 2013 and 2018.[28] One reason is that some insurers balk at the prospect of tying their insurance coverage decisions to standards.

In California, for example, regulators have tried to replicate the Wildfire Partners program, but it has proven difficult to recruit insurance companies to participate. According to Dave Jones, the insurance companies worry that if a resilience standard emerges and they lend their support to it, then regulators might eventually force them to cover everyone who meets the standard. They prefer to decide whom to cover and whom to drop on a case-by-case basis, and they do not welcome the loss of discretion, even if the outcome would be good for the state and the homeowner.[29] Yet the standards-led approach remains promising as a science-based, market-driven way to promote resilience on a large scale. Regulators and state legislatures may eventually have to get involved to encourage insurers to join the effort.

As our review of the four markets suggests, climate risk disclosure can act as a powerful and transformative force, one that can propel climate resilience at a systemic level. But transparency can also prove highly disruptive. In all four markets, fear of disruption and its consequences has led different groups to throw sand into the gears to delay the day of reckoning. But that day is coming, as the impacts of climate change will eventually expose the companies, municipalities, homes, and countries that run the highest risk.

As that process unfolds, if there have been no efforts to prepare, investors, banks, and insurance companies could panic and pull back indiscriminately from parts of their respective markets. Investors and financial-services companies could opt to dump shares, bonds, property, and insurance customers, motivated more by panic than by an accurate understanding of actual risks, causing widespread economic harm. A smarter way forward is for financial regulators to work with the private sector to lay out a gradual but firm path toward mandatory disclosure and clear standards. This will give people time to put resilience measures in place and teach the market how to make more discerning choices between those who manage climate risk prudently and those who prefer to ignore it.

PRESCRIPTIONS AND PROVOCATIONS

- Financial regulators, in close consultation with the private sector, should put in place mandatory requirements for disclosing the climate risk of publicly listed companies.
- Business leaders should lead a process to develop an authoritative protocol that enables companies to understand and report material climate risks in a comparable, standardized way.

- The US Congress should reform the National Flood Insurance Program by phasing out subsidies over an appropriate transition period, excluding new development in floodplains from coverage, and terminating coverage for properties that have flooded repeatedly, all while providing adequate assistance to affected homeowners, especially low-income households.
- Insurers in the United States should offer premium reductions to property-owners and communities that invest in verifiable risk-reduction measures and meet higher building standards.
- State governments should provide property-tax breaks to owners who retrofit their homes or small businesses for greater resilience and comply with the latest model building standards. Federal, state, and local government grants should be provided to homeowners for incorporating approved resilience measures.

PART II

TOOLS FOR THE DECISION-MAKER

[4]

FIND BETTER WAYS TO PAY
FOR RESILIENCE

On the third floor of the building that houses the US Treasury Department hang portraits of the seventy-six former secretaries of the treasury, starting with Alexander Hamilton. The men in all the paintings are wearing jackets, some even waistcoats, except for one. The portrait of Hank Paulson (Figure 4), who served under President George W. Bush, is of a man standing in his shirtsleeves, shirt slightly untucked, sleeves rolled up, hands in his pockets, a look of amused puzzlement on his face. Paulson is unique not only for rejecting a formal pose. History will best remember him for having engineered the Bush Administration's response to the 2008 global financial crisis and for persuading Congress to approve the controversial Troubled Asset Relief Program. Depending on your political persuasion, the program either bailed out crooked banks at taxpayers' expense and should never have happened, or it heroically averted another Great Depression, or both.

Less well known is that Paulson is one of a small handful of prominent Republicans who favor aggressive action to combat climate change. In 2014, long after leaving the Treasury Department, Paulson was asked about his personal strategy for talking to fellow Republicans skeptical of efforts to cut greenhouse-gas emissions.

Figure 4 Portrait of Hank Paulson, by Aaron Shikler, 2010. Source: US Department of the Treasury.

Paulson knew only too well that bank bailouts—the use of public funds to rescue troubled financial institutions from bankruptcy—are deeply unpopular with the American public. Yet, he explained, no one who wants a future in politics can afford to turn his or her back on a disaster-stricken community. When a major natural disaster strikes, the government steps in and pays for a large share of the uninsured losses. As climate change makes extreme weather more frequent and natural disasters more severe, Paulson continued, the losses will stack up. Government debt will get bigger and bigger. "Climate bailouts," as Paulson termed the use of public funds to help affected communities recover and rebuild after a natural disaster, will become a regular fixture of national life. So if Republicans care about limited government, Paulson concluded, they should care about controlling climate change before it results in never-ending climate bailouts.

Recent history suggests Paulson was right. Between 2005 and 2008, Congress appropriated almost $130 billion to pay for natural-disaster damages, caused mostly by Hurricanes Katrina, Rita, and Wilma.[1] After Superstorm Sandy struck, in 2012, the government paid out over $50 billion. And following devastating wildfires and Hurricanes Harvey, Maria, and Irma, in 2017, Congress made available almost $140 billion in emergency funding. Congress borrowed most of this money, adding to the growing national debt.

Even for the largest economy in the world, ever-larger climate bailouts are not a responsible solution to handling present and future climate impacts. They will cut deeper and deeper into vital areas of public spending, such as infrastructure, education, and health care. They will feed a spiral of borrowing, leading to higher financing costs for the government and higher taxes. Escalating climate bailouts will accelerate the declining fiscal health of the country, which will make policy trade-offs a lot tougher for the next generation of Americans. For this reason, the US Government Accountability Office (GAO), a trusted government watchdog that works for Congress, identified climate change as a "significant financial risk to the federal government."[2] In 2013, the GAO added climate change to its annual list of issues that constitute the highest fiscal risk to the US government. As of this writing, climate change is still on the list. For other nations, especially those in the world's poorer regions, borrowing on this scale is not an option. The traditional approaches, hoping that foreign aid will flow sufficiently quickly and in adequate amounts or otherwise leaving needs unaddressed, are not good options either.

In the coming years, governments everywhere, including in the United States, will have to raise unprecedented amounts of money to cope with the impacts of climate change. Precise estimates are hard to find, but one review of the literature suggests that countries should already be spending between 0.67 percent and 1.25 percent of their annual gross domestic product (GDP) on resilience.[3] Globally, that means hundreds of billions of dollars per year, and currently, countries may be underspending on resilience by as much as 70 percent.[4]

How can communities raise the money needed, and how can they do so while keeping the financial strain as low as possible? They can fund resilience the old-fashioned way, through tax revenue, borrowing, and buying reinsurance. For developing countries, securing more international assistance will be necessary. But governments must also deploy new ideas, including those we discuss in this chapter—setting up special reserve funds, using value capture, raising funds from carbon taxes and cap-and-trade mechanisms, and issuing green and catastrophe bonds. Climate bailouts, even for the richest nations, are not a smart way to grapple with the effects of climate change. We can and must do better.

TAXING OUR WAY TO RESILIENCE

Miami Mayor Tomás Regalado was not always an advocate for climate resilience. A grizzled Republican with over twenty years in Miami electoral politics, he focused his political message on practical, pocketbook issues. "Do I have a vision?" he repeated when the press asked him the question in 2013: "Keep taxes down. Reduce the size of government. Fix the potholes. Fix the streets. Pick up the garbage."[5] According to the *Miami Herald*, it was Regalado's son, an underwater photographer named José, who made a point of sitting down with his father at four or five o'clock in the morning and, over cups of Cuban coffee, sharing articles and exchanging ideas about climate change. By 2017, the mayor had come to terms with the fact that his city had become emblematic of the impacts of climate change.

With little time left in his last term in office, Regalado threw his political weight behind an uncharacteristic initiative—raising taxes to pay for resilience. Miami would borrow $400 million from the market by issuing bonds, and through a referendum, city residents would agree to increase their taxes to pay it back, with interest. Authorities would dedicate almost half the money to upgrading storm drains, installing flood pumps, and building or strengthening seawalls. There was no effort to hide the bonds' climate-resilience objectives. The initiative was known as the Miami Forever Bond. "The city eventually has to deal with this," Regalado said, referring to the growing problem of flooded streets. "And the only way the city can do that is with the bonds."[6] Voters approved the referendum, with 55 percent of them supporting the measure.

Taxes are the most obvious and old-fashioned way to finance investments in climate resilience. Governments can increase general taxes, such as income or sales taxes, or they can introduce taxes targeted for resilience measures. In 2016, for example, voters in the San Francisco Bay Area approved by a wide margin an annual, $12 property tax to restore the bay's natural wetlands. This green infrastructure should provide significant flood protection benefits by acting as a buffer as sea levels continue to rise. A year later, San Francisco voters overwhelmingly approved, by an 80 percent margin, an initiative to issue $425 million in bonds to strengthen the Embarcadero seawall. The project will protect $100 billion worth of property from seismic risk, as well as from the rising waters. Voters in Harris County, Texas, whose seat is the city of Houston, voted for a tax increase in 2018 to finance a $2.5 billion "flood bond," the proceeds of which will pay for flood mapping, an improved flood early-warning system, and infrastructure to expedite drainage.

Some developing countries and small island states are doing the same thing. The climate-vulnerable Pacific island of Fiji adopted in 2017 an Environment and Climate Adaptation Levy (ECAL), a 10 percent tax on a wide range of items, from plastic bags to restaurant meals to movie tickets. Fiji's government expects that ECAL revenues will exceed $47 million per year, or about 4 percent of Fiji's total tax revenue. It has committed to using the money for climate-resilience projects.[7]

In many places, from the United Kingdom to several US states, governments are using taxpayer resources to capitalize green banks. These are financial institutions with special mandates to finance projects that generate environmental benefits. So far, green banks have focused most of their climate-related lending on efforts to cut emissions, and very little money has gone to building resilience. This is a missed opportunity; green bank capital could go a long way in helping communities prepare for climate impacts.

As a strategy to raise money for climate resilience, taxes have limits. Raising taxes is politically challenging, even if politicians promise to dedicate the revenue to cope with climate change impacts. Some taxes may be regressive, hitting the poor disproportionately. Also, one-off tax measures and bond issues will not deliver the sustained funding needed to meet the resilience challenge. Policymakers will need to combine taxes with other strategies for raising cash to pay for resilience.

PAYING OUR FAIR SHARE

New York City's subway acts as the heart and arteries that keep the city running. But the system has not kept up with the city's growth and needs renovation and expansion. In their search for ways to finance the revitalization, city and state officials have proposed one mechanism, known as "value capture," to raise funds for new subway-related projects. This strategy can potentially enable communities to invest in climate resilience projects as well.

The concept is straightforward. Everyone should pay his or her "fair share" of the cost of new infrastructure that brings very localized benefits to their communities. In the case of a subway line, for instance, proximity to a station improves property values; the closer a building is located to a subway station, the more convenient it is for residents and workers to access, and the more valuable the property becomes. In Manhattan's main business corridors, for example, proximity to the subway adds an estimated $4.58 per square foot to the value of commercial property.[8] So, the argument runs, it makes sense for those who derive direct benefits from proximity to the subway to share some of the costs above and beyond what they contribute in general taxes. This approach is best suited to paying for renovations to a particular station or for the extension of a subway line to reach a particular community. General upgrades to the entire system, which benefit everyone, are more appropriately funded from general tax revenue.

How could New York put value capture to work? New York authorities have proposed conducting before-and-after assessments of neighborhood property values whenever the city plans to build a new transportation project costing more than $100 million, such as a subway line extension. Authorities would calculate the difference in property values before and after the project is completed, as well as the difference in property-tax revenue. The government would then transfer most of the estimated gain in tax revenue to the transit agency to pay for the improvement. A similar approach has been considered to fund improvements to the Paris Metro.[9]

Communities could apply similar strategies to climate resilience projects. For example, the value of a property protected from storm surge by a new seawall or natural barrier typically should be higher than the value of a similarly situated property that does not have such protection. Authorities can estimate the value difference between comparable properties with protection and those without it. Some of that difference can then be "captured" through increased taxes on the benefiting properties. This is a big undertaking; making the calculations is not easy. They will also be hotly contested by property owners and they raise issues of equity. But given the potential benefits of this approach, it's an experiment worth trying.

PAYING WITH CARBON

In 2005, in a remarkable feat of climate leadership, European leaders launched the European Union Emissions Trading System (EU ETS). A cornerstone of Europe's efforts to fight climate change, the EU ETS applies the concept of "cap and trade" to a significant swath of industry. Like other cap-and-trade systems, it works like this: The government limits the total amount of permissible carbon emissions and then issues permits that allow permit holders to legally emit a certain amount of carbon. Businesses can buy and sell permits in a special market. Companies that can reduce their emissions relatively cheaply will choose to do so rather than to buy permits, whereas companies for which cutting emissions is expensive will prefer to buy permits instead. As the government reduces the total amount of permissible emissions over time, the price of permits will increase, pushing more and more emitters to cut emissions. When the system works as intended, it reduces overall emissions and generates revenue from the sale of permits in the process.

Since the purpose of the system is to fight climate change, it is fitting that the government should use at least some of the proceeds from the sale of the permits to prepare for its impacts. But this has not been the case so far. Between 2013 and 2015, the EU ETS raised about €12 billion (about $14 billion). The EU spent most of the money on emissions-reduction efforts; only a miniscule amount was spent on climate resilience in Europe and beyond.[10] In the United States, California's cap-and-trade mechanism generated about $4.5 billion between 2012 and 2016. Some of the activities that were funded with this money benefited resilience indirectly. But the state has yet to designate any of this money for resilience activities.[11] The Regional Greenhouse Gas Initiative (RGGI) is a regional cap-and-trade system run by a group of northeastern US states. It has raised at least $2.6 billion. Of all the RGGI states, only Delaware appears to have used a portion of its share to build resilience, in this case for coastal protection and flood prevention.[12]

Governments aren't planning to use money raised through carbon taxes for resilience either. Carbon taxes are fees charged by the government and paid by the emitters of greenhouse gases; they are typically levied on each ton of carbon emitted. In 2018, voters in Washington state considered one of the first referendums on a carbon tax to be held in the United States. One of the most progressive climate efforts in the country, Initiative 1631 would have raised an estimated $2.2 billion in the first five years. Yet less than one-twentieth of the revenues would have gone to assist communities in buffering against climate impacts.[13] In any event, the referendum failed to get the necessary votes to pass. As the world adopts more carbon taxes and cap-and-trade mechanisms to cut emissions, putting aside some of those revenues for resilience will generate significant funding streams that can enable communities to prepare for the impacts of climate change.

GREEN BONDS FOR RESILIENCE

In 2007, Aldo Romani was head of the investor relations team at the European Investment Bank (EIB), the world's largest multilateral development bank. Nestled in Luxembourg, on a campus of well-tended gardens and contemporary art sculptures, the bank serves as the public policy bank for European Union member states. For an EU member country that needs financing for an EU policy priority, the EIB is one of the first stops. At the time, the EU had set out climate-related targets for reducing greenhouse-gas emissions, boosting renewables, and increasing energy efficiency. But implementing the plan, known as "20-20-20," after the percentage targets set in each category, required funding.

The challenge of raising money for the initiative fell partly on Romani's shoulders. The bank could easily issue bonds to raise funds from the usual suspects, but Romani hoped to attract new investors. His answer was to market a new line of securities that would appeal to an emerging class of "green" investors. The securities were called Climate Awareness Bonds, and they promised investors that their money would be invested only in green projects, such as renewable energy and energy efficiency. Back then, selling bonds expressly for climate activities was still a somewhat edgy, untested strategy, so the bank initially issued bonds worth a relatively modest €600 million. The early buyers were mostly European investors interested in "socially responsible" or "sustainable" investments and, despite the bonds' unmemorable name, they sold well.

A few months later, the World Bank issued its own green bond, this one denominated in Swedish kroner and similarly targeted at European investors. As with the Climate Awareness Bonds, investors quickly scooped up the World Bank bonds. Yet mainstream investors continued to regard green bonds as something of a curiosity. They figured these securities were best suited for do-gooders required to fulfill environmental mandates, not for traditional investors in the hard-nosed business of maximizing investment returns.

But something interesting happened in the decade after Romani's bond sale. Heavyweights, including the Chinese government and Apple Corporation, issued their own green bonds. Industry groups agreed on common definitions and standards to govern the acceptable use of monies raised from green bonds. A cottage industry of third-party validators appeared to reassure investors that their green-bond investments were going only to authorized, green activities. In 2017, green-bond issuances exceeded $150 billion, and Wall Street started to take this formerly niche market seriously.

Despite the bonds' considerable success, however, the issuers of green bonds have not typically used the proceeds to build resilience. Understandably, issuers have directed the vast majority of green-bond proceeds to pay for reducing greenhouse-gas emissions. Still, a few green bonds with resilience components have emerged.

The city of Cape Town, South Africa, in its efforts to avoid running out of water, issued a bond to pay for technology to build resilience against water stress. The Brazilian pulp, paper, and packaging company Klabin issued a green bond to pay for, among other things, efforts to reduce the company's exposure to wildfires. Yet proceeds from the sale of green bonds could do a lot more to build resilience. Indeed, "resilience bonds" should eventually emerge as subcategory of green bonds. They should become a well-understood standard product that investors interested in supporting resilience will want to buy.

THE PARIS GRAND BARGAIN

In December of 2015, at a military airport outside Paris, most of the world's nations approved the historic Paris Agreement, creating a truly global framework for tackling the climate challenge for the first time. At the heart of the agreement lay a grand bargain: developing countries agreed to join developed countries in reducing emissions, and developed countries agreed to raise money to help developing countries pay for emissions cutting and resilience. Although the developed nations never explicitly recognized it, the financial arrangement was an attempt to grapple with one of the central injustices of climate change, namely, that the states least able to cope with climate change are also the countries least responsible for causing the problem. This is especially true of small island states and most of the world's low-income regions, whose contribution to the stock of greenhouse gasses in the atmosphere pales in comparison to that of the industrialized world and the largest emerging economies.

How much money did the parties to the Paris Agreement settle on? Developed countries promised to raise $100 billion per year by

2020. That includes direct financial assistance from rich-country governments to be channeled through aid agencies, multilateral banks, and other institutions, as well as money raised from the private sector. That commitment continues until 2025. Thereafter, the number will increase to a to-be-determined level above $100 billion. Developed nations hope they can pressure China—now the world's largest greenhouse-gas emitter—to join them in contributing money as part of an expanded financial commitment.

For poorer nations, this money is vital to pay for resilience. Richer economies are doing a decent job so far in delivering on the $100 billion, but that amount is not nearly enough to cover the growing needs. The money must be spread among dozens of countries and must stretch to cover both greenhouse-gas-reduction and resilience efforts. So far, governments are spending only about a quarter of the money on resilience. Many poor nations find it difficult to access the money in the first place because of slow domestic and international bureaucracies and cumbersome requirements. Yet despite these challenges, international climate finance is a resource worth tapping into, especially as rich countries, and perhaps the developing countries that emit a lot of carbon, consider contributing more resources.

In addition to raising funds to pay for resilience before disaster strikes, governments also need to raise money to pay for recovery and rebuilding after the damage has been done. There are several tools decision-makers can use.

FORCING OURSELVES TO SAVE

In 1995, a double disaster hit Mexico. The country suffered a financial crisis that pushed the economy into a deep recession. Then in

October, Hurricanes Opal and Roxanne pummeled Mexico's Gulf Coast and Yucatan Peninsula, killing hundreds of people and causing extensive damage. The government scrambled to make money available for disaster response, but the economic crisis hampered these efforts. Resolving to do better, the finance ministry commissioned its head of budget policy, an economist named Fausto Hernández Trillo, to find new financial options for responding to disasters.

Trillo and his team designed the Natural Disaster Fund, better known by its Spanish acronym, FONDEN, to pay for postdisaster relief and reconstruction. Importantly, the Mexican Congress enacted legislation requiring the government to commit at least 0.4 percent of each year's federal budget to FONDEN. In a typical year, that has amounted to some $800 million. By embedding this mandatory contribution in the law, Mexico has created a stable source of funding that it can use to respond to disasters and invest in resilience projects. The law forces the government to save, no matter which political party is in power.

Other countries have followed in Mexico's footsteps. Mozambique passed a law requiring at least 0.1 percent of its national budget be used to fund the country's Disaster Risk Management Fund. Similarly, the Marshall Islands, the Philippines, Kenya, Jamaica, and Guatemala—nations that are all highly vulnerable to climate change impacts—have all set up national reserve funds. Of course, governments can also use reserve funds to pay for investments in resilience, not just for recovery.

These funds face risks, however. Politicians may find the temptation to divert money from the disaster fund to pay for unrelated priorities or pet projects irresistible. To counter this temptation, the trick is to protect the money. Strong rules on use of the money, extensive information disclosure requirements, and clear lines of responsibility within government agencies can help ensure that the money goes only to resilience-related uses. This will make it harder for politicians to raid the rainy-day fund, and the money will be there when it's needed.

The United States has a Disaster Relief Fund, which is the federal government's main source to pay for disaster relief and recovery. When the fund's balance runs too low, Congress refills its coffers, but unlike funds in other places, this is not automatic. In a welcome development, Congress passed legislation in 2018 requiring that six percent of the post-disaster assistance provided by the US Federal Emergency Management Agency (FEMA) should go into the agency's Pre-Disaster Mitigation Fund. This should ensure that billions of additional dollars are used to build resilience before disaster strikes.

SHIFTING RISK TO THE PRIVATE MARKET

In January of 2017, the head of FEMA, Craig Fugate, closed a deal with an industry few people know about: reinsurance. Reinsurance companies insure other insurers. For a price, reinsurance companies sell insurance coverage to primary insurance companies and other clients who seek protection against some of the world's most extreme risks. Reinsurers can afford to deal in this type of risk because they have large capital reserves and globally diversified insurance and investment portfolios. In theory, this means that they will always have money to cover claims because the chance of multiple large-scale catastrophes happening at the same time has historically been very low.

For years, Fugate had been concerned about the financial health of the federally run National Flood Insurance Program (NFIP). As we saw in chapter 3, the NFIP's financially unsustainable arrangements have saddled the program with growing debts. Fugate worried that future disasters would put the NFIP deeper in the red. To deal with this challenge, FEMA purchased a $1 billion insurance policy from global reinsurance companies. If a single flood event cost FEMA more than $4 billion during the year, reinsurers would pay a quarter of the losses.

Events soon demonstrated the prudence of Fugate's move. Eight months after the reinsurance purchase, Hurricane Harvey devastated the Gulf Coast, causing massive losses and triggering $8 billion in NFIP payouts to homeowners whose properties had been damaged or destroyed. FEMA's reinsurance policy paid out the full $1 billion, and in 2018, the agency purchased even more reinsurance, this time $1.5 billion of coverage. Similarly, the United Kingdom's Flood Re (see chapter 3) program that assists homeowners in managing flood risk, has bought billions of pounds worth of reinsurance.

Governments around the world have also turned to reinsurance markets to raise money quickly after natural disasters. In the developing world, countries in Africa and the Caribbean have banded together in regional pools that allow them to buy reinsurance more cheaply than they could on their own. During the Obama administration, the United States helped extend similar pools to Pacific islands and nations in Central America. Through these risk pools, governments buy so-called parametric insurance policies, which pay out automatically and quickly when an event of a certain magnitude occurs. There is no need to wait for damage assessments, which can take weeks or months. In this way, parametric policies provide fast cash to respond to droughts, hurricanes, earthquakes, and floods. This matters, because experience shows that the quicker and more effective the initial response to a disaster, the lower the long-term economic damage. Reinsurance, including parametric products, will remain a key tool to enable governments and businesses to raise money for coping with climate impacts.

BETTING ON CATASTROPHE

In addition to reinsurance, governments and businesses have turned to catastrophe, or "cat," bonds to secure cash quickly after a disaster. A cat bond is another way for a company or government to purchase catastrophic risk insurance. Instead of buying the protection from a reinsurance company, the government or company buys it from scores of investors.

A cat bond works like this: Assume a business or government wants the ability to respond quickly in the event of a disaster or just wants more protection to cover catastrophe losses. So it issues a cat bond, which is essentially a contract with the investors who buy the bond. The company or government pays the investors interest every month and, in exchange, the investors agree that if a certain predefined natural disaster happens during a specific time period, the company or government automatically gets the money from the sale of the bonds. Investors walk away empty-handed, except for any interest payments they have received. Cat-bond investors are essentially placing a bet that catastrophe will not strike.

After Superstorm Sandy, for example, New York City's Metropolitan Transit Authority (MTA) spent hundreds of millions of dollars repairing the subway system. To secure this money, the authority pulled resources away from other vital priorities. MTA officials wanted a better way to raise emergency funds, so they issued a $200 million cat bond in 2013. The authority agreed to pay investors 13.5 percent interest, but if at any point during the bond's life, independently managed tidal gauges at different points around the city showed storm surge exceeding a certain height (at Battery Park, on the city's southernmost tip, it was 8.5 feet, or 2.6 meters), the bond would automatically trigger payment of the bond proceeds to the MTA.

New York City was not alone in turning to cat bonds. To add to its reinsurance coverage, FEMA issued its first cat bond in 2018 for $275 million to protect the NFIP's finances. Meanwhile, Mexico uses cat bonds to protect its national reserve fund, FONDEN, from large losses. Mexico, Colombia, Peru, and Chile jointly launched a $1 billion cat bond to cover earthquakes.

As of 2019, the cat bond market exceeded $30 billion, and it continues to grow as more governments and businesses seek financial protection from climate-driven extreme events. It's not hard to see why. The issuers of cat bonds like the product because it provides an alternative to reinsurance that is sometimes more cost-effective and transparent. Investors like the high interest rates that cat bonds typically pay. They also appreciate how a cat bond diversifies their portfolio, since the value of a cat bond has little to do with the economic factors and trends that affect other investments, for example economic growth, interest rates, and the price of oil.

As former Treasury Secretary Hank Paulson warned his Republican colleagues, climate bailouts will remain politically irresistible. As long as the US government continues this reactive approach, the country's fiscal health will continue to suffer, leading to serious economic consequences for future generations. In countries where climate bailouts are less feasible, the alternatives will be even worse. Climate-driven extremes will force developing nations to rely even more heavily on unstable sources of foreign aid or to simply go without adequate support. Communities and businesses must do better, and that means deploying both old-fashioned approaches and financial innovations to raise funding not only for disaster relief, but also for resilience.

PRESCRIPTIONS AND PROVOCATIONS

- The federal government and states should direct meaningful portions of revenues from cap-and-trade systems and carbon taxes to investments in resilience; green banks should also adopt resilience as a core part of their mission.
- Green-bond standard-setting organizations should develop definitions and standards to govern the acceptable use of proceeds from the sale of resilience bonds and actively publicize new offerings of resilience bonds.
- State and local governments and the private sector should pilot value-capture methodologies to finance resilient infrastructure.
- State governments should pursue forming risk pools to purchase parametric insurance products or issuing cat bonds to access cash quickly in the aftermath of disasters; the federal government should continue to protect the NFIP through the use of resinsurance and/or cat bonds.

[5]

GET THE DATA AND MAKE
THEM USABLE

In the early days of the Cold War, as the mistrust between the United States and Soviet Union gave way to open confrontation, the US military flew regular spy missions over Soviet territory. Flying above the range of Soviet antiaircraft weapons, the spy planes would collect valuable intelligence. But the military had a problem. Bad weather sometimes prevented the planes from successfully completing the mission. To ensure that its pilots would have good visibility when they arrived in enemy airspace, the US military needed to forecast conditions over the Soviet Union. Collecting data through conventional means, such as weather balloons, was not going to work; the Soviets would immediately shoot down any foreign objects within range of their antiaircraft weapons. So the American generals turned to a think tank called the RAND Corporation for advice. In 1951, RAND issued a secret report that would not be declassified for over forty years.[1] It proposed a novel approach to forecasting that involved "weather observations made by means of a television camera placed in an unmanned vehicle flying above the normal range of defense weapons." In other words, a weather satellite.

Almost a decade later, RAND's proposal came to life when the US government launched the Television and InfraRed Observation

Satellite, or TIROS-1—the first weather satellite in history (Figure 5). The satellite had two television cameras that took a distorted, black-and-white picture of a small part of the Earth's surface every ten seconds. Each camera linked to a magnetic tape recorder that could store up to thirty-two photographs when the satellite's signal was out of the range of stations on the ground. A power failure permanently disabled the cameras after just a few weeks in operation, ending the mission. Still, in its brief life, TIROS-1 beamed back some 20,000 images, including one of a cyclone above New Zealand, proving that weather observation from space was possible.

Since the days when TIROS-1 represented the cutting-edge of technology, the world's capacity to collect and analyze climate and

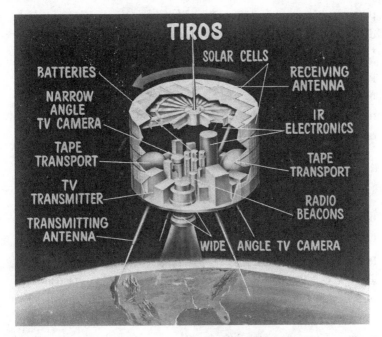

Figure 5 Artist's rendering of the instruments aboard TIROS-1 satellite.
Source: National Aeronautics and Space Administration.

weather data has exploded. The problem now is that we face a paradox. At no point in history have these data been more plentiful and predictive models more powerful. Yet many of the people who need them most lack both access to the right data and the means to make them useful for decision-making.

Consider, for example, Perdido Beach, a tiny, climate-vulnerable town on the coast of Alabama in the United States. Given its location, the community suffers from high exposure to hurricanes and sea-level rise. In 2012, Hurricane Isaac washed away portions of the beachfront. Two years later, floods washed out some roads in Perdido Beach, trapping residents inside their homes. Coastal protection is critical for the town's survival, but how much infrastructure should it build? What kind? At what cost? And what is the best placement to maximize protection? In order to make these critical decisions, the community needs data and modeling, but even in the United States, small communities lack the resources necessary to plan for resilience. Speaking to federal authorities, Perdido Beach Mayor Patsy Parker explained her predicament: "I'm just a part-time mayor in a small town. I don't have a big planning staff, grant writers, or any resources. So how can I even know the size of the threats we are facing—and what can I do to protect the people of my town?"[2] In most developing countries, the problem is even more dire.

As climate change advances and its impacts become clearer, more communities in the United States will need deeper insight into the future, both immediate and distant. Decision-makers will require information to make high-impact, hard-to-reverse decisions about water, agriculture, and where and how to build infrastructure in a world experiencing climate change. They will demand data to calculate the likelihood of catastrophic events and to figure out how best to plan and pay for defenses. They must model the projected evolution of droughts, heatwaves, and wildfires, so they can help people get out of harm's way. And they will need data to make disaster-relief operations more effective. In this chapter, we describe the data paradox and offer a few ideas on how to escape it.

THE GOLDEN AGE OF CLIMATE AND WEATHER DATA

Since the 1960s, technology has made possible what can only be described as a revolution in weather and climate data. Consider satellites. One of the most modern is the GOES-16 (Geostationary Operational Environmental Satellite), launched by the US government in 2016. It provides a "full disk" image of the Earth centered on the Western Hemisphere every 15 minutes and an image of the continental United States every five minutes. It can monitor hurricanes, wildfires, and other severe weather events as often as every 30 seconds and take pictures with sixteen spectral bands, meaning that it can "see" dust, haze, smoke, fog, ice, snow, changes in vegetation, and moisture levels in the atmosphere. When Hurricane Maria knocked out radar systems in Puerto Rico in 2017, GOES-16 enabled meteorologists to monitor the massive storm as it made its way across the Caribbean. Satellites have come a long way since TIROS-1's wobbly, black-and-white transmission.

The data revolution no longer depends just on expensive, government-owned satellites. After working at the US National Aeronautics and Space Administration (NASA), British-born physicist Will Marshall launched the company Planet Labs (now simply called Planet). His big idea was to piggyback on advances in miniaturization to build tiny satellites, or "cubesats." No bigger than a shoebox and costing a small fraction of the amount of traditional satellites, cubesats are launched dozens at a time in "flocks." Once they are in orbit, they spread out to cover the planet. Although they are not substitutes for GOES-type satellites, each Planet satellite can collect 10,000 images, covering 770,000 square miles (2 million km^2) per day.[3] Today, a flock of cubesats that is already in orbit provides daily images of the Earth's entire landmass.

Despite the notoriety that drones have gained for their controversial role in targeted assassinations, when it comes to climate resilience, drones provide a valuable service. Packed with sensors, digital radios, and powerful microprocessors, drones enable farmers to see their fields at high resolution with multispectral bands, which can reveal potential irrigation problems, water stress, or pest or fungal infestations. Drones can detect healthy and stressed plants more accurately than the naked eye, allowing farmers to focus their efforts on trouble spots. Drones can do this cheaply and regularly, providing the farmer with time-series animations that reveal how the field is responding to weather and farmer interventions. Heavy-duty drones, such as those used by NASA's Hurricane and Severe Storm Sentinel Mission, can get close to hurricanes and collect data on the forces that drive them. After a disaster has struck, drones can survey the damage quickly, allowing for instant identification of areas most in need of assistance, which may speed up the recovery process.

Ground and sea-based sensors also play a role in the data revolution by providing superlocalized information. As Hurricane Sandy approached the northeastern United States in 2012, US government scientists scrambled to deploy water-level and barometric pressure sensors in over two hundred locations along the Atlantic coast, from Virginia to Maine.[4] Data from these devices proved crucial to emergency managers and insurers, who required the most precise information about the extent and costs of the damage. Startup companies have developed ground-based, wireless sensing devices that are solar-powered and independent of the cellular network, so that they can keep collecting and transmitting data even if electricity and telecoms grids go down.

Mobile phones can turn anyone into an environmental data-collection agent. Thousands of volunteers armed with cell phones now regularly collect information about the impacts of extreme weather events and disseminate it on social media. These efforts are also making it possible to map previously unmapped parts of the world, so that communities can better understand flood and heatwave risks. People carrying cell phones can even generate useful information unwittingly. Call detail records, which time stamp and pinpoint the location every time a person makes or receives a mobile phone call, can help emergency-management agencies understand, in near-real time, how a population responds to disaster alerts. They also enable emergency managers to track population movements after a natural disaster.

Thanks to all this technology, the volume of data collected today is impressive. The US National Oceanic and Atmospheric Administration (NOAA) alone collects 20 terabytes of environmental data from its satellites and other sources every day. That's the daily equivalent of about 8,300 ninety-minute Netflix videos.[5] As of 2018, the satellite company Planet was collecting about six terabytes a day from its satellites—about 2,500 Netflix videos a day.[6] We are awash in data. But how do we make it useful for decision-making and accessible to the people who need it?

MODELING THE FUTURE

To render all the data we collect useful, weather data agencies and companies must clean, process, and package it so it can be digested by decision-makers. Machine learning, and artificial intelligence (AI) more generally, can help us sort through mountains of data rapidly. Programmers can teach computers to "read" the massive inflow of data and look for relevant patterns. Computers can do this much faster and more accurately than humans can, and they work around the clock, without coffee breaks. Machine-learning programs can also modify themselves as they interact with the data, making automatic adjustments to get better at their task. They can also share what they learn with other machines, multiplying the speed of improvement across the network.

But even when decision-makers can access and crunch the right data, they need insights about the future, not just about the past and present. For that, decision-makers need models that can inform forecasts. Models are simplified simulations of how something works. The stronger our understanding of the science behind climatic events, and the larger the volume of historical data we can feed into the models, the better the models become at forecasting the future. And the more powerful the models become, the greater the ability of government and business leaders to make informed decisions about climate resilience.

Climate and weather models are getting better all the time. Take weather forecasts. Today's five-day forecasts are as good as two-day forecasts were twenty-five years ago. Five-day cyclone forecasts are now the global standard.[7] Cloud computing has helped make this possible. By buying massive online storage capacity, software, and computing power and renting it out to thousands of users, "cloud" providers enable their customers to store and analyze large quantities of data at a fraction of what it would cost if they bought their own equipment. This switch has helped advance the development and use of climate and catastrophe models.

Few groups should be as motivated to model the future of extreme weather as insurance and reinsurance companies and catastrophe bond investors (see chapter 4). These groups have lots of money riding on this information. Depending on how much damage hurricanes, earthquakes, floods, windstorms, and wildfires cause (or don't cause), insurers and investors can either earn or lose millions or, increasingly, billions. Catastrophe ("cat") models can deliver an informational edge by helping them understand the likelihood that a disaster will strike a given place, the amount of property that is vulnerable to the disaster, and the value of the damage that could result. The better the models, the more accurately insurers and cat-bond markets can set the price for protection, and the better the chance that those being insured are not underpaying (or overpaying) for coverage. Of course, cat models have other uses. For instance, urban planners can use them to make better decisions about where to put infrastructure, and emergency managers can employ them to understand which parts of a country or region are at greatest risk.

Blind spots still bedevil cat models, however. The models are good, for example, at forecasting wind damage from a storm. But they have much more trouble computing flood damage, which depends on many more factors, such as elevation and topography, that determine where the water flows after it hits the ground. For this reason, all the major commercial cat models in the past have consistently underestimated the size of insured losses caused by major hurricanes affecting the United States.[8] Wildfire models are getting more and more accurate at forecasting which areas are at greatest risk for fire, and once a fire starts, how it might behave. But as with hurricanes, the models break down when it comes to the biggest fires. Fires that get big enough and hot enough to generate their own complex wind patterns, including "firenados," are, so far, too complex to model accurately.[9]

When it comes to modeling climate impacts, one important challenge is "downscaling." Scientists have generated reams of information regarding climate impacts on global, regional, and national scales. But information to forecast how climate impacts will unfold in a small geographic area or corporate facility remains harder and more expensive to obtain. Yet this information is critical to making good resilience decisions. Everyone from local officials to supply-chain risk managers, urban planners, and emergency-management agencies needs to understand how climate change-related events might impact specific factories, farms, ports, buildings, and critical pieces of infrastructure. If the models spit out a "pixel size" that is too large, then the information will be too general to be actionable.

Entrepreneurs are taking on the challenge. Startups in Silicon Valley and other places are hoping to bring super-localized forecasting to government and to the private sector. Highly granular modeling is becoming available, especially in the most lucrative markets of the United States and Europe. One Concern, a Silicon Valley startup, has launched products that promise to forecast flood levels and other perils with "asset level" precision—meaning at the level of individual buildings and industrial facilities.[10] In Florida, an environmental lawyer and a climate scientist have teamed up to help clients protect themselves from flooding related to sea-level rise using proprietary climate-modeling technology. The firm, Coastal Risk, provides clients with maps of projected tidal and storm-surge flooding at a resolution of up to one square meter (10.76 square feet). As the business case for climate-predictive analytics becomes more and more compelling, entrepreneurs will compete with growing intensity to give their clients an informational edge to build resilience.

WHO GETS THE DATA, AND
AT WHAT PRICE?

While climate and weather data have become more plentiful, and models more powerful, the question is who can access them and at what price? As climate change advances and the need to build resilience becomes a matter of burning urgency, this question will become political and ethical, not just economic. In the United States, policymakers have long debated whether government or private sector should take the leading role in the provision of weather and climate data, and they have gone back and forth on the issue several times.

Consider the story of Landsat, a series of US government weather satellites that have been gathering data since 1972, providing an unbroken stream of images of the Earth and documenting natural and human-caused changes to the planet in ever-greater detail. During the 1970s and 1980s, the government considered privatizing American weather satellites and transferring Landsat data to private companies for distribution. In 1984, Congress prohibited the privatization of the government's satellites, but it did pass a law allowing the transfer of Landsat data to a private vendor under an exclusive license.[11] That vendor could then sell the data for a profit. The company, called the Earth Observation Satellite Company (EOSAT), promptly raised prices for Landsat data and focused on developing sophisticated data products for the high end of the market.[12] After the price hike, many customers for basic data products, including academic institutions, could no longer afford it.

Less than a decade later, Congress reversed itself, concluding that "the cost of Landsat data has impeded the use of such data for scientific purposes, such as for global environmental change research, as well as for other public sector applications."[13] Landsat data distribution shifted back to the government, which charged only the cost of fulfilling users' requests. But the pressures to privatize continued. Another attempt to turn Landsat data over to the private sector came in 2002, unsuccessfully, and in 2012, when the Department of Interior asked the Landsat Advisory Group to consider privatization. The group advised against it. Today, Landsat data is freely and openly available to all, at a cost of essentially zero. But as the economic value of weather and climate data continues to rise, the debate over privatization will likely return.

Limited access to catastrophe models, not just to the underlying data, also poses concerns. Cat models remain expensive. A handful of cat modeling companies has dominated the market for years, and some blame this situation for slowing innovation and keeping costs high.[14] The proprietary models of those companies are closed "black boxes." Insurers and other clients pay handsomely to use these tools under a commercial license, but they only own the outputs of the model, not the model itself. Also, insurers have historically had a hard time comparing the cost and quality of the models produced by the various providers.

Developing countries have an especially hard time getting access to models that will enable them to understand the impacts of climate change. Poorer regions of the world are often not lucrative places for insurers and reinsurers to do business. As a result, modeling companies have not spent resources modeling natural disasters there, and so climate risk in those countries remains poorly understood. That, in turn, makes it less likely that people there will want to buy insurance. In an effort to break this vicious cycle, government and academic institutions have developed free, open-source models, but they are still few and far between. The most comprehensive inventory available currently lists 240 cat models, but only 9 percent are freely available, and some provide only limited functionality.[15]

The government's role in providing basic public goods in the form of climate data and information is more important than ever. Public agencies should provide a bedrock of free, taxpayer-funded climate and weather data that all levels of government can use to inform public policy, to protect public assets, and to build the resilience of poor and vulnerable people. It should also continue to fund the basic climate science that underlies climate models. Meanwhile, private companies can specialize in customized, proprietary products. Governments typically lack the necessary expertise to do the latter, and the private sector usually innovates more quickly than government agencies. Also, if companies need highly downscaled data and models to safeguard their own facilities and supply chains, it seems fair that they pay for that.

Realizing that government needs to play a strong role in providing climate information, countries around the world have started to strengthen their weather and climate agencies. Uruguay, for example, transformed its meteorological agency from a unit within the Defense Ministry into a separate entity, giving it greater independence and flexibility. In 2016, the government of Uruguay also established a national-level platform that serves as a one-stop source of climate data and information. It also provides tools to help farmers make climate-informed decisions.[16] Another example is the India Meteorological Department, which replaced an obsolete weather-prediction model with a customized version of a system developed by the US National Centers for Environmental Prediction.[17] The new system is designed to enable India's meteorologists to issue more granular and accurate forecasts in a country where understanding monsoon rain patterns and extreme heat events is vital to lives and livelihoods.

Meanwhile, some US government officials have given serious thought to how best to strengthen the nation's climate information system. In 2015, the US Government Accountability Office (GAO), the nonpartisan watchdog agency that serves Congress, published the results of an illuminating investigation.[18] Climate information in the United States, the report found, "exists in an uncoordinated confederation of networks and institutions."[19] The federal government's climate information, the report went on to say, "is fragmented across many individual agencies that use the information in different ways to meet their respective missions . . . [D]ecision makers are vastly underserved by the current ad hoc collection of federal climate information services."[20] The GAO also studied the national climate information institutions of the Netherlands, the United Kingdom, and Germany, all of which the watchdog regarded as more effective than the American approach.

What to do about the US government's fragmented system? Creating a new centralized agency to manage climate information may seem tempting. Yet the GAO inquiry rightly concluded that a more decentralized system would best serve the United States because no single agency has the expertise required to provide a full menu of climate information services. At the same time, a strengthened US approach should give one federal agency the authority to push other government departments to share information with the public in "customer-friendly" formats. Above all, the federal government should put a system in place that can harness the vast amount of climate and weather information it collects and make it available in accurate, reliable, and easily accessible ways across multiple, complementary platforms. Any such effort must remain laser focused on ensuring that those who most need the information can readily put it to immediate use.

Meanwhile, cities are also building data infrastructure to help them manage climate impacts. New Orleans, for instance, built one of the best data hubs in the country. Its Data Center brings together information from multiple sources to track the city's demographic and economic recovery after shocks such as Hurricane Katrina. New Orleans has also created geospatial maps that enable decision-makers to visualize multiple layers of information about the urban environment simultaneously. These maps enable officials and others to pinpoint city blocks where temperatures may be dangerously high and to identify the most strategic locations for placing green infrastructure. Officials in Norfolk, Virginia have taken notice, and they are studying New Orleans's achievements. They hope to establish a data center of their own.

LOST WITHOUT TRANSLATION

Noted psychologist Daniel Kahneman has observed, "No one ever made a decision because of a number. They need a story."[21] With climate information, the challenge is how to make the data speak to decision-makers and the public. The experts who generate the data sometimes forget the critical nature of this translation challenge. What good are the best models powered by the best data if the users fail to draw actionable insights from them?

Translating climate and weather data can take different forms. Some approaches harness the power of visuals. One example is Aqueduct, an open-source online tool developed by the World Resources Institute (WRI), a nonprofit research organization.[22] Aqueduct takes data and modeled projections of water stress and generates colorized maps. The user can see how his or her community is expected to fare in terms of water availability as far as 2040.

Areas expected to remain rich in water are shown in a soothing cream color, while water-stressed areas become progressively redder. High-risk places appear in bright red, and those at "extremely high risk" show up in a deep, burnt maroon.

Owens Corning, a company based in the Rust Belt town of Toledo, Ohio, put this tool to use. The firm's 20,000 employees produce insulation, roofing, and fiberglass composites. The work requires lots of water to cool the machinery and materials used in high-temperature manufacturing processes. Owens Corning has facilities in areas of Mexico, Spain, China, and the United States that will struggle to meet water demand. The company layered a map of its global facilities on top of Aqueduct's colorized maps, making it possible to identify which facilities faced the highest risk of water stress, now and in future decades.[23] Guided by this analysis, Owens Corning ranked its facilities worldwide to determine where to focus its resilience budget.[24]

Sometimes translation is all about tailoring the message to the audience and delivering it through the right messenger. In Colombia, for example, climate experts noticed that farmers there were not thinking enough about how climate change might affect their crops. Climate-impact scientist Julian Ramirez-Villegas recalled, "[A] lot of people were aware of the importance of the climate, but didn't know what to do about it." "After going to the field, we realized that Colombian farmers were planting based on what happened last year . . . With the amount of climate variability that we have here in Colombia, particularly rainfall, that's a recipe for disaster."[25] Farmers needed information not just about the weather over the next few days, but also about conditions several months into the future.

A nonprofit organization dedicated to helping farmers in developing countries cope with climate change impacts took on the challenge. Based in southwestern Colombia, the International Center for Tropical Agriculture (CIAT), employs over three hundred scientists and boasts a depository with seeds for 37,000 varieties of beans. To help Colombian farmers understand future risk, CIAT generated forecasts using crop models and big-data analytics. The forecasts provided the farmers with very localized climate projections extending several months into the future. CIAT aimed to help the farmers answer three practical questions. Where should they plant, given the forecast? When is the best time to do so? And which crop variety should they put in the ground?

The next challenge was how to get the information to the people who needed it most. CIAT worked with farmer organizations to build tools for relaying the forecasts to their member farmers. But many small-scale farmers lived in remote areas and didn't belong to farmers' organizations. To reach them, CIAT turned to radio, television, and text messages. CIAT calculates that 300,000 farmers are now receiving climate information thanks to this initiative. The organization wants to take the effort global. "Imagine if we could implement similar systems in sub-Saharan Africa or South East Asia," says Ramirez-Villegas. "There are potentially millions and millions that could benefit."[26]

In the United States, the Obama administration launched an initiative called Climate Services for Resilient Development, along with a slew of other efforts aimed at helping make climate information more useful for developing nations, including Colombia. Another novel initiative launched by the administration was the Climate Resilience Toolkit, a web portal offering tools and information designed to help Americans inside and outside government understand climate risks and their implications for different regions of the country better. In a report issued in the administration's final days, the White House recognized that despite these efforts, a lot more needed to be done in this area.[27]

We live in a golden age of climate and weather data and analytical and modeling power. Yet we experience an acute shortage of insights to help drive resilience. Breaking out of this paradox involves collecting and downscaling data and ensuring that it is accessible. It demands addressing the weaknesses of the current catastrophe models and translating climate information in ways decision-makers can use. And it requires that we put in place effective weather and climate agencies and foster a cadre of translators who can demystify climate information.

PRESCRIPTIONS AND PROVOCATIONS

- The federal government should rethink the organization of the US climate data system, drawing on lessons from other countries, to ensure that climate and weather data from across the government remains free, openly sourced, and disseminated in user-friendly ways.

- Cities and states should consider creating data centers that collect and analyze local demographic and environmental data that are useful for resilience planning.

- Insurers and reinsurers should invest in catastrophe models that better reflect future climate risk and work with regulators to ensure that insurance premiums reflect that risk. The federal government, in partnership with the private and non-profit sectors, should expand the availability of open-source, free catastrophe models.

- The federal government, in partnership with the private sector, should create a cadre of climate-science translators to help decision-makers in state and local governments evaluate climate risks, develop resilience strategies, and access federal funding.

[6]

WORK WITH HUMAN NATURE

In 2017, the city of Cape Town, South Africa's second most-important economic center, began to run out of water. Experiencing a drought of historic proportions, officials in the city of nearly four million people watched with trepidation as water levels in key reservoirs declined at alarming rates. Most worried was Patricia de Lille, a former laboratory technician and labor-union leader who served as Cape Town's mayor. "We are in a very serious crisis and we know there are still some customers who are acting as if our resources are not under strain," she warned in February 2017.[1] In October 2017, de Lille said that "[i]f consumption is not reduced to the required levels of 500 million liters (132 million gallons) of collective use per day, we are looking at about March 2018 when supply of municipal water would not be available."[2] The fateful date soon became more precise: March 13, 2018. The countdown began to what came to be known as Day Zero—the day when, for the first time in modern history, the taps of a major city would run dry.

To avert the crisis, de Lille and her team rolled out every tool available. The government raised water rates, imposed fines on delinquent water users, tightened enforcement, and expanded leak detection and repair. It rushed to bring new infrastructure on line to increase water supply, including new desalination plants. The city imposed a water-use target of 50 liters (~30 gallons) per person per day—enough for drinking, cooking, hygiene, doing dishes and laundry, a ninety-second shower, and one toilet flush.

But as late as January 2018, only about half of the city's residents were complying with water restrictions.[3] So the mayor complemented the carrots and sticks already in place with tools drawn from behavioral science. In addition to all its other measures, her government experimented with "nudging," that is, applying techniques designed to prod behavior in a predictable direction, but which by themselves do not forbid options or significantly change economic incentives. The government turned to a study commissioned some years before by the South Africa Water Research Commission that suggested significant additional water savings could occur if customers received notice of how their excessive water use compared to their neighbors' water use.[4] The government decided to harness the power of peer pressure—naming and shaming, as well as recognizing achievement—to change behavior.

Cape Town authorities applied this concept on a large scale. The city launched a web-based water map, which allowed anyone to see the water usage of individual properties across the entire city, based on the latest meter readings. Properties using less than 1,600 liters (6,000 gallons) showed up on the screen with a bright green dot; those consuming more than that were shown with a dark green dot. The goal was to turn the city bright green. The city also publicly identified the streets where the one hundred largest water users resided. Alongside this, Cape Town authorities distributed information and "tips" on how to reduce water consumption.

The concept of Day Zero itself proved key to the government's nudging strategy. Day Zero made the threat real—particularly to the largest water users and wealthy owners of water-guzzling lawns and pools. The city launched a web-based "water dashboard," showing weekly data on the water levels in reservoirs, the city's collective water consumption, and how consumption compared to the government's targets. The authorities also set up a Facebook page illustrating some of the measures that would be put in place should Day Zero arrive, including vivid descriptions of the hardships of water rationing, the locations of water distribution points, and business closures. The pain these measures would impose was not lost on well-heeled Cape Town residents, who started referring to this as "the crazy page." "Day Zero," one Cape Town resident and environmental advocate told us, "scared the shit out of middle-class white people."

In the end, Day Zero did not arrive. High-income households cut water demand by 80 percent, and low-income families reduced it by 40 percent.[5] Many of the demand-reduction measures, including the behavioral interventions, remain in place and are becoming the new normal. To be sure, not every intervention worked as expected, and some of the measures, such as the water map, remain deeply unpopular. It is not yet clear whether Cape Town avoided disaster or merely postponed it; that will depend on the city's long-term planning. But the combined measures achieved massive cuts in demand in a short amount of time, and that has officials from water-stressed regions around the world, from California to São Paulo, trying to learn from Cape Town's story.

Cape Town's water crisis illustrates a larger phenomenon. Human beings are not psychologically well-equipped to prepare for the impacts of climate change. We are not good at dealing with dangers we have trouble picturing in our minds, and we often succumb to excessive optimism. We are reluctant to pay short-term costs that are certain in exchange for future, uncertain benefits. Given the enormity of the climate resilience challenge, we are at risk of feeling overwhelmed and therefore paralyzed by the scope of the problem. If we are going to build resilience to climate change successfully, we are going to have to work around some of these cognitive limitations. Human nature is difficult, if not impossible to change, so it is best to deploy a variety of approaches and "nudges" that work with human nature, not against it.

GRAPPLING WITH HUMAN BEHAVIOR

Never before have policymakers paid as much attention to the scientific study of human behavior as they do today. Over the past twenty years, psychologists, neuroscientists, and economists have joined forces to understand in more rigorous ways how humans make decisions. Behavioral science has rapidly captured the attention of decision-makers in politics and business. In 2002, American-Israeli psychologist Daniel Kahneman received the Nobel Prize in economics for his insights into human judgment and decision-making. His 2011 book, *Thinking, Fast and Slow,* has become a bestseller and is frequently found on the bedside tables of influential people.

In 2010, British Prime Minister David Cameron created the Behavioural Insights Team, nicknamed "the Nudge Unit" inside his office. He gave it two years to prove that behavioral science could improve public policy and deliver a tenfold return on its costs.[6] The unit proved its worth by demonstrating how modest interventions could, for example, boost the rate at which taxpayers paid back taxes and the number of people who chose to insulate their attics, cutting energy waste. The unit has since reincorporated as a public-service company and today supports public-service organizations globally. In 2015, President Obama created the Social and Behavioral Sciences Team at the White House and ordered all executive agencies to develop strategies for applying behavioral science to programs and recruit relevant experts.[7] At least a dozen countries have integrated the insights of behavioral science into their operations.[8] So far, however, nudge units have not paid enough attention to helping communities cope with climate impacts. It's time to put these insights to work for resilience.

Behavioral science suggests that certain biases regularly impede good decision-making. A long catalogue of these exists. Here, the focus is only on a few that are most relevant for climate resilience. A key one is *availability bias*, or the tendency to judge an event based on how easily we can call to mind a relevant example. This bias can lead us to underestimate the likelihood of events we have never seen before. In the context of a changing climate, this is obviously a problem, since we are constantly confronted with new extremes.

Consider the surprise expressed by politicians, news commentators, and homeowners at the never-before-seen trajectory of Hurricane Sandy, which drove it into New York and New Jersey from the Caribbean in 2012, or the record-breaking scale of California fires in recent years, or the fact that 2017 was the first time in recorded history that three Category 4 hurricanes made landfall in the United States in a single year. Climate change will create conditions for which humans have no collective memory. To adapt to climate change, we are going to have to visualize possible futures, even those that seem quite unfamiliar.

Another stumbling block is *optimism bias*, or the tendency to overestimate the likelihood that we will experience good events and underestimate the likelihood that bad events will befall us. Many people believe they will be safe from extreme climate events not because they can't imagine disasters happening, but because they don't imagine disasters will happen to them. This tendency shows up clearly in polling data.

In a 2018 public opinion survey, over 70 percent of respondents said that global warming will cause harm to plants, animals, and future generations.[9] Sixty percent expect that harm will also come to people in the United States, and half said the harm will extend to their communities and even their own families. But only 42 percent think they will be personally harmed by climate change. Likewise, millions of people let optimism get the best of them when it comes to property insurance. Once memories of a disaster fade, the number of people buying coverage declines as customers let their fire or flood insurance policies lapse. Many can't help but think that they will somehow escape through skill or luck, even when others are afflicted.

Another cognitive bias worth highlighting is *loss aversion*, which is the tendency to give greater weight to losses in comparison to gains. Loss aversion comes from the perception that a loss hurts, say, twice as badly as a gain feels good, even when the loss and gain are objectively of the same size. This is a well-documented phenomenon in the investment world. An investor might insist on holding on to an investment that is losing value, even though the economically rational thing to do would be to cut losses and sell immediately. Yet the mental pain of realizing the loss keeps the investor from dumping the asset, which only makes his or her financial situation worse. With regard to resilience, loss aversion bias can mean that communities avoid making investments today that carry short-term costs, even if those investments will protect them in the long run from bigger climate-related losses.

INSTITUTIONALIZING IMAGINATION (COUNTERING AVAILABILITY BIAS)

After the attacks of September 11, 2001, the US government set up a commission to study the event and draw lessons on how to guard against similar tragedies. In its final report, the commission famously identified "failure of imagination" as a factor in the government's inability to foresee and prevent the massacres. It concluded that "at least some government agencies were concerned about the hijacking danger and had speculated about various scenarios. The challenge was to flesh out and test those scenarios, then figure out a way to turn a scenario into constructive action."[10]

How to address this failure to take preventive action? "Imagination is not a gift usually associated with bureaucracies," the commission concluded dryly. "It is therefore crucial to find a way of routinizing, even bureaucratizing, the exercise of imagination."[11] This sounds like a contradiction in terms. We think of imagination as something that belongs to the realm of unstructured, serendipitous thinking, not to the regimented world of bureaucracy.

Part of the answer lies in making the exercise of imagination a regular, even mandatory practice. This is what scenario planning is all about. It's not about predicting the future, but about considering what different futures might look like regardless of how likely or unlikely they may seem. The goal is to expand the range of worlds and events decision-makers can visualize, especially when these are outside their direct experience. If these worlds become more mentally "available" to leaders and the public, they will be able to grapple more effectively with these possible futures.

Scenario planning focused on climate impacts is already being put to use. In 2017, more than a dozen globally active banks, many of them household names, got together to ask a single question. How might climate change affect customers and their capacity to repay loans in the years and decades to come?[12] The banks modeled several scenarios involving average global-warming increases of 2°C (3.6°F) and 4°C (7°F) above preindustrial levels. What they found was sobering but also helpful.

The Canadian banking giant TD Bank looked at how climate change might affect the electric utilities to which it lends money. It subjected a sample of its customers to simulations of a warmer climate and concluded that, under all scenarios, a majority of the utilities would see their creditworthiness deteriorate because of climate change impacts. Another major bank, which decided to remain anonymous, analyzed a sample of its clients in the agriculture sector and found that almost all would see significant declines in revenue and credit downgrades in a 2°C warmer world. In a 4°C warmer world, the decline would double in severity. Interestingly, most of the financial damage in both banks' simulations resulted from slow-onset impacts, such as water stress and rising average temperatures, not from sudden extreme events, such as hurricanes.

Scenario planning is not just for banks. The private-sector-led Task Force on Climate-Related Financial Disclosures (see chapter 3) has recommended that all companies undertake climate change scenario analysis and disclose the results. Another example is the Obama administration's work with state and municipal leaders to use scenario planning to prepare for local climate threats. In Texas, the White House and the Federal Emergency Management Administration (FEMA) worked with Houston to imagine a hurricane hitting the city, eerily foreshadowing Harvey's devastation several years later.

In 2014, the White House and FEMA also teamed up with local leaders in Norfolk to analyze climate-impact projections, discuss their implications for the city, and propose short-term actions to begin addressing long-term risks. Participants in the exercise considered risk in two time horizons—one referred to as "our children's time," which ran until 2044 and another called "our grandchildren's time," which ran until 2084. FEMA's planning division deliberately framed the exercise in terms of time horizons that would feel personal and real to participants; that's why the planners referenced cycles of family life. The exercise produced a range of thoughtful recommendations, some of which have since been implemented, including revising the city's building code to require resilient construction.

Another way of institutionalizing imagination is through design competitions. In 2013, the US government launched an initiative called Rebuild by Design. The program offered over a billion dollars to support innovative projects that leveraged climate resilience strategies to rebuild areas hit by Superstorm Sandy. Conceived as a competition, Rebuild by Design hoped to unlock the creativity needed to build for a different future. One of the first projects to be awarded funding was called "The Big U" (Figure 6). Its architects proposed constructing a belt of hardened, flood-protection infrastructure all around lower Manhattan, but this infrastructure would not look like a barrier separating the community from the waterfront. Instead, the architects designed it to double as a structure for public recreation, sightseeing, environmental education, and even farming. "We put designers at the center of [the program] because design is a discipline that teaches you to imagine things that people haven't seen," recalled Shaun Donovan, Obama's Housing and Urban Development Secretary.[13]

Figure 6 Rendering of the Big U in Manhattan, New York City. Source: ©
Rebuild by Design.

TAKING OFF THE ROSE-COLORED
GLASSES (COUNTERING OPTIMISM BIAS)

At 10:11a.m. on August 28, 2005, as Hurricane Katrina was closing
in on Louisiana, the local office of the National Weather Service
(NWS) issued a public warning that screamed, in characteristic capital letters:

DEVASTATING DAMAGE EXPECTED . . . ALL WOOD FRAMED LOW
RISING APARTMENT BUILDINGS WILL BE DESTROYED . . . THE
MAJORITY OF INDUSTRIAL BUILDINGS WILL BECOME NON
FUNCTIONAL . . . AIRBORNE DEBRIS WILL BE WIDESPREAD . . . PER-
SONS . . . EXPOSED TO THE WINDS WILL FACE CERTAIN DEATH IF

STRUCK . . . WATER SHORTAGES WILL MAKE HUMAN SUFFERING IN-
CREDIBLE BY MODERN STANDARDS.[14]

Subsequent analyses of evacuation during Katrina have suggested
that this terrifying official statement convinced people to evacuate
and saved lives.[15] The message shook people out of their optimism
bias and got them to take off their rose-colored glasses. In the complex world of emergency warnings, public officials must work hard
to nudge people out of their optimism bias and get them to evacuate. Sending a scary message is one way to do it, but experience
shows that authorities should use this approach carefully. Getting
people to respond appropriately in the face of danger is more difficult than it seems.

Sometimes, scary messages can backfire. In one revealing experiment, conducted in Florida's Miami-Dade County and published in
2016, researchers created a fictional storm ("Hurricane Julia") and
provided local residents with information about the storm.[16] They
tested different warning messages on different groups of residents
to see how they would react. Not surprisingly, the group that received a "scary" message (IF . . . YOU STAY IN THE AREA, YOU MAY
DIE) reported a higher desire to evacuate than those who received
a less scary message: THERE WILL BE STORM SURGE OF 4 FEET OR
HIGHER ALONG COASTAL AREAS. But the "scary" message led many
participants to regard the information as overblown and the official
source as unreliable.

The experiment also showed warnings that cite probabilities can
backfire, too. For example, when some residents received a probabilistic warning (THERE IS A 55 PERCENT CHANCE THAT THE EYE OF
THE HURRICANE WILL MAKE LANDFALL IN MIAMI-DADE COUNTY),
they wrongly associated the 55 percent likelihood not only with hurricane landfall, but also storm surge and other impacts. More people

decided to play the odds and stay put, even when the probabilistic warning was accompanied by the scary message (YOU MAY DIE).

At the same time, the experiment found that when residents received an actionable warning (IF YOU LIVE IN AN AREA AT RISK FROM STORM SURGE OR FLOODING, EVACUATION IS THE MOST EFFECTIVE WAY TO PROTECT YOURSELF AND YOUR FAMILY), more people decided to evacuate, and fewer felt the information was overblown or unreliable.

All this suggests that public officials face tricky dilemmas. Do they omit discussion of probabilities because people may misunderstand them? Do they risk putting people in harm's way if they don't use scary messaging, even if it undermines their credibility in future emergencies? Behavioral science seems to be saying that optimism bias can be managed with the right messages. Scaring people can prove effective but must be done sparingly. Giving the public clear information about the risks and providing them with actionable recommendations can nudge desirable behavior. As we move into a world where extreme weather may become a permanent feature of life, resilience will mean finding the best ways to get people to protect themselves instead of making foolhardy wagers with their lives and property. That will involve undertaking much more research into how people respond to different messages.

RETURN THE MONEY OR BUILD THE FLOODWALL? (COUNTERING LOSS AVERSION)

In the fall of 2012, Superstorm Sandy flooded 80 percent of the city of Hoboken, New Jersey. Best known as Frank Sinatra's birthplace, the city was founded on an island surrounded by marshland, making it highly prone to flooding. During Sandy, flood waters swamped all the city's electrical substations, casting its 53,000 residents into darkness and immersing streets in a dangerous mix of water, toxic chemicals, and raw sewage. After the waters receded and the city restored some semblance of normality, Hoboken Mayor Dawn Zimmer made it a top priority to protect her city from future storms.

Under Zimmer's leadership, Hoboken applied for and received $230 million in funding from Rebuild by Design. Hoboken's plan called for a floodwall and structures designed to delay flood waters, a retention system that would hold up to a million gallons of water, and a pump station to discharge the retained water. But before it could receive the money, the city would have to go through an extensive process of public consultation and analysis to narrow down the design options. Here Zimmer's floodwall and other design features ran into a different kind of wall—a wall of public opposition, some of which reflected residents' loss aversion bias.

Even though Sandy's destruction had not yet faded from memory, many Hoboken residents staunchly opposed the design options, fearing that their property values would take an immediate hit if the city built unsightly flood-protection structures close to their homes. The plan provided long-term benefits for the city and its residents by making them safer from flooding, but the minds of many residents focused only on their immediate losses due to devalued properties. Zimmer recalls an early meeting with residents: "I arrived at the meeting with two hundred infuriated people, who literally were screaming at me to give back the money."[17]

No doubt helped by her background in crisis communications, Zimmer committed herself to finding a way forward. That way forward involved highlighting the present and future benefits of the project for residents. Zimmer organized dozens of neighborhood meetings and community gatherings and visited concerned residents. She pushed the engineers to come up with alternative design options. They shifted the location of some structures and, echoing New York's "Big U" proposal, added parks, benches, murals, and green walls so that the protective infrastructure would offer Hoboken's residents amenities. The process enabled people to focus on the present and future benefits of the new infrastructure, rather than solely on their immediate losses. Zimmer secured public backing for the project, and Hoboken got the grant. The following year, the Metropolitan Waterfront Alliance recognized Zimmer as a "Hero of the Harbor" for her work in making Hoboken a national model for preparedness.

THE POWER OF WANTING TO FIT IN

The science of nudging leans heavily on the idea that the framing of choices can influence what people choose. One way to foster certain types of resilient behavior is to capitalize on the fact that people want their peers to respect and admire them. Consider the findings of a large-scale experiment in Atlanta, Georgia, the results of which were published in 2011.[18] Researchers divided about 100,000 water customers of the Cobb County Water System into three groups (with a fourth control group). One group received a two-sided sheet offering tips on how to reduce water usage and how to get more information. A second group got the tip sheet plus a "soft social norm" message in the form of an official letter making the case for water conservation: "Reducing our water consumption today is important for preserving our environment and our economy for future generations." The third group received a "strong social norm" message. This was a letter reporting a resident's total water consumption and comparing it to his or her neighbors' average consumption for the same period. It also contained a message saying, for example, "You consumed more water than 73 percent of your Cobb County neighbors."

The experiment found that the strong social-norm message resulted in the largest water savings. If all households in the experiment had received the strong message, they would have saved the county an estimated 186 million gallons (700 million L) of water—the equivalent of cutting the water service to some 5,100 households. The highest water users proved to be the most responsive to the message. Indeed, high users reduced water consumption by almost 6 percent, while low users averaged only a 3 percent reduction. Perhaps this difference simply reflects the fact that high-user households, with their pools and lawns, had more room to cut than low users, whose water needs were more basic. Disappointingly, though, the study also found that the effect of the measure wears off with time. Three months after the strong social norm letter led to big reductions in water use, about a third of the water savings disappeared as some users returned to their old ways.

Back in Cape Town, researchers also discovered the power of peer comparison. Around the time of Day Zero, a team of researchers tested a suite of nudging techniques to identify what worked best at inducing residents to save water.[19] They found that of all the techniques, the most effective was advertising to customers that the city government would publicly recognize the biggest water savers. (Those who preferred not to be "outed" could opt out.) On average, using that approach alone resulted in water-usage reductions of almost 2 percent, or nearly 132 gallons (500 L) per household per month. Harnessing the power of social recognition for climate action, and for resilience in particular, can be a useful nudging tool, though it must be combined with other policies to yield lasting results.

Behavioral science holds no silver bullet for meeting the challenge of climate resilience, but the examples in this chapter show that the field offers useful and actionable insights. For that reason, countries that have established "nudge units" should focus significantly more attention on climate resilience. Those that have not set up such units should strongly consider doing so. The World Bank has assembled its own behavioral science team and has identified some tips for how to establish a nudge unit.[20] These include enlisting a champion within the government, securing a two- to three-year commitment to give the unit time to show results, and focusing first on the low-hanging fruit—issues on which behavioral science theories can be tested quickly, easily, and at little cost.

Establishing *climate* nudge units dedicated to emissions reduction and climate resilience efforts could improve decision-making. The climate nudge units could focus on such challenges as institutionalizing imagination and road-testing messages that shape the public's perception of climate risk. The units could also conduct large-scale experiments, share their results, and learn from one another. Leaders in business and at all levels of government should actively participate in the effort. The resulting nudges can provide communities with another tool that can be brought to bear on the challenge of building resilience.

PRESCRIPTIONS AND PROVOCATIONS

- Governments should apply behavioral science insights to relevant policies, programs, and operations, where such insights are likely to advance significantly climate resilience.
- Large companies should integrate regular climate-risk scenario analysis into key planning processes, including

assessments of climate impacts on their supply chains and continuity of business operations.

- The federal government should support a research program to road test the effectiveness of different types of communications and warnings regarding climate and weather-related threats.

- A climate nudge unit should be established at the federal level, with a clear mandate and resources to study, design, pilot, and evaluate behavioral science initiatives that promote climate resilience.

PART III

THE UPENDERS

HARDEN THE HEALTH CARE SYSTEM, AND MAKE IT SMARTER

On December 28, 2015, the *New York Times* ran a story that sent shivers down the spine of public health workers around the globe. The story described the appearance of a virus in Brazil that was being spread by mosquitos. Despite its catchy name—Zika—the virus carries devastating consequences for babies in the womb. Though not typically life-threatening, Zika is known to cause microcephaly, an incurable form of brain damage in newborns whose mothers have been infected with the virus during pregnancy. Fetuses had displayed such catastrophic brain damage that some prominent obstetricians in Brazil were advising their patients not to get pregnant.

It was the third paragraph of the article that particularly grabbed the attention of White House officials. It quoted the Centers for Disease Control and Prevention (CDC) warning that the number of Zika cases imported to the United States would "likely increase and may result in local spread of the virus in some areas of the United States."[1] Before the *New York Times* report appeared, few on the National Security Council staff at the White House had been tracking the Zika virus. But the CDC announcement changed all that. Zika quickly came to dominate daily life for many at the White House, just as its staff was winding down the administration's response to the threatened spread of the deadly Ebola epidemic. In January 2016, the Zika virus likely appeared in the United States for the first time; a baby was born in Hawaii with the virus and microcephaly to a mother who had lived in Brazil.

As concern mounted, the authorities swung into action. The White House convened a task force to oversee the response, and the CDC created a Zika Pregnancy Registry to track potential cases and medical outcomes. The following month, the World Health Organization declared the outbreak an international health emergency. The White House sent a request to Congress for $1.8 billion in emergency funding to fight the spread of the disease, though the proposal quickly got ensnared in politics. In August, the CDC instructed pregnant women and their partners to stay away from one of Miami's trendiest neighborhoods, which had emerged as the epicenter of Zika in the country. It was the first time the CDC had ever advised the public not to travel to a community in the continental United States to avoid catching an infectious disease. Within weeks, Florida began aggressively spraying pesticides to stop the outbreak and continued to do so despite public opposition. By the end of the year, the CDC's Zika registry had received reports from forty-four states of pregnant women who had tested positive for Zika infection. Of the reported births from mothers with confirmed cases of Zika, an alarming 10 percent of the babies had birth defects.[2]

Zika's growing reach serves as a cautionary tale. Health experts expect mosquitos to migrate to new areas as temperatures rise, further spreading not only the Zika virus, but also viruses such as Dengue, West Nile, and Chikungunya.[3] Other vector-borne diseases spread through the bites of ticks, fleas, and mosquitos, may also make inroads in regions these pests have previously found inhospitable.

Climate change will likely introduce new stresses to human health, and not only from the geographical expansion of vector-borne diseases. Because climate change affects basic elements central to human survival—air, food, and water—more people in more places will face public health threats.[4] Extreme heat is a killer. Climate change is also expected to affect access to fresh water and increase the likelihood of human and animal exposure to water-borne diseases, including those spread through bacteria and toxins produced by harmful algae. Food security will be affected as a result of both lowered crop yields and reductions in the nutritional value of certain foods. Indeed, climate change may well affect the entire food chain, from production to storage, distribution, and consumption. At the same time, the trauma of climate change-related events will stress mental health, especially among the elderly, the young, and those already suffering from mental illness. Extreme climate events can overwhelm health care facilities and dramatically impair the delivery of health care services.

Despite the clear and present danger climate change poses to public health, the efforts of the medical profession and public health authorities to date have not matched the urgency of the problem. Medical facilities are not being designed, built, and retrofitted to withstand climate impacts. Because we lack sufficient capacity to track, monitor, and warn the public about the spread of disease in a warming world, we find ourselves in a reactive mode, struggling to keep up. Nurses, doctors, and public health officials are often unaware of how climate change upends their understanding of public health threats and what to do about the growing risks.

While the challenge is daunting, there are certain things we can do to prepare. Here we focus on a relatively narrow but important element of the response—the reliability and effectiveness of systems delivering medical services. Communities, governments, and global institutions must develop ways to provide health care in a changing climate. This calls for hardening the infrastructure supporting the health care system so that it can withstand climate impacts and continue to deliver services. Health care systems must also be made smarter, which means training health care providers about climate-exacerbated health conditions, putting in place predictive and early-warning systems, and finding more effective ways to protect and attend to the most vulnerable people.

HARDENING THE HEALTH CARE SYSTEM AGAINST CLIMATE CHANGE

A health care system must perform reliably, but climate change-exacerbated events will threaten that reliability. When Superstorm Sandy struck New York City in October 2012, about 6,500 patients had to be evacuated from six hospitals and thirty-one residential care facilities when these buildings lost power.[5] Backup diesel generators quickly ran into problems because builders had installed critical elements such as fuel pumps on lower floors that flooded. At Bellevue, the city's oldest hospital, workers formed a "bucket brigade" to move fuel from the flooded basement's fuel storage to the generator, thirteen flights up.[6] At New York University's Langone Medical Center, the loss of power forced medical personnel to evacuate newborn babies from the neonatal intensive care unit, some tethered to battery-powered respirators.[7]

Auxiliary health care providers also struggled to provide services. In the greater metropolitan area of New Jersey and New York, patients dependent on home nursing care, electric-powered devices, and refrigerated medicine suddenly lost access to these critical services. As professionals later picked apart the medical response to Sandy, they found that seemingly unrelated disruptions unleashed by the storm further compromised health care delivery and access. Entire neighborhoods were cut off from medical facilities when debris-clogged roads and public-transportation systems shut down. Critical medical supplies, such as portable oxygen, ran short as delivery systems stopped.[8] Many health care workers were unable to get to work and faced the added burden of worrying about the safety of their own families. In total, Sandy caused billions of dollars in damages to hospitals and health care organizations in New York City alone, not counting the hard-to-measure costs to patients of delayed or suspended access to health care.

But even without a Sandy-caliber storm, emergency health services can be disrupted by climate-related events. In flood-prone Norfolk, for example, chronic flooding can make streets impassable, delaying the delivery of potentially life-saving care. In 2017, nearly half of residents living in Norfolk and the surrounding communities reported being unable to drive out of their neighborhoods at least once in the past year because of flooding during high tide.[9]

Ignoring the risks of climate change impacts to health care infrastructure costs lives. Consider what happened when Hurricane Irma swept across Florida in 2017. The storm knocked out power to over six million residents, including the Rehabilitation Center at Hollywood Hills, a 152-bed nursing home that housed elderly patients. Loss of power caused the air-conditioning system to fail. Temperatures began to rise, climbing to more than 100°F (38°C) on the second floor. Three days passed before the facility was completely evacuated. By then, a dozen people had died. Criminal authorities investigated the deaths as homicides, and the nursing home lost its license. In the wake of the disaster, the governor of

Florida signed legislation that requires nursing homes to have adequate back-up power systems—a "no more" moment for those caring for the elderly. A year after Irma pummeled Florida, the US Congress took the effort nationwide, directing training for local governments and utilities on ways to prepare for loss of power in medical facilities and nursing homes during extreme weather events.

Pockets of climate-resilient health care—facilities designed to provide health care even in dire conditions—have begun to emerge in some parts of the world. Take, for example, Georgetown Hospital in St. Vincent and the Grenadines, a small Caribbean archipelago that is no stranger to hurricanes. Originally constructed in the 1980s, Georgetown Hospital provides care for close to 10,000 people. In 2013, it launched the Georgetown Smart Hospital Project aimed at building and retrofitting the facility to make it resilient to hurricanes. Retrofits included new water-storage systems, roof upgrades, natural ventilation and lighting, installation of a generator, and energy-efficiency improvements, all at a cost of under $350,000.[10]

Another example is the Texas Medical Center (TMC), a 2-square-mile (5 km^2) medical district in Houston, Texas. It houses over sixty medical institutions, making it the largest medical complex in the world. TMC employs over a hundred thousand people and records millions of patient visits each year, generating an annual revenue comparable to the gross domestic product of a small country. In 2001, Tropical Storm Allison hit the complex hard, halting almost all of the center's operations and inflicting millions of dollars in damages. Vowing to be better prepared the next time, TMC invested more than $50 million in resilience measures. It installed watertight flood doors, widened drainage culverts, elevated electrical equipment and water pump systems, installed stormwater tanks to capture run-off, created a flood-alert system, increased training, and built a berm around the McGovern Medical School that stands 1 foot (30.5 cm) above the 500-hundred-year floodplain (the area that has a 0.2 percent chance of flooding in any given year).[11] When Hurricane Harvey hit in 2017, TMC closed its flood gates and went about the business of caring for patients. Every hospital on TMC's campus remained open but one. Babies were delivered and emergency brain surgery was performed. Because medical personnel knew the hospital would remain open, they did everything they could to be there to help; some even kayaked through the flood waters to get to the TMC.[12]

It's not just major infrastructure investments that can save lives. Preventive steps that carry little or no cost can make a significant difference. Take, for example, what medical personnel did in Ahmedabad, the largest city in the Indian state of Gujarat. In 2010, a heatwave struck the city. Admissions to the neonatal intensive care unit at one of the city's un-air-conditioned hospitals spiked dramatically. The maternity ward was housed on the hospital's top floor under a dark tar roof, and hospital administrators knew they had to find cooler quarters for the newborns. They relocated the maternity ward to the cooler ground floor. The next time temperatures hit 108°F (42°C), that simple move could reduce heat-related admissions to the neonatal intensive care unit by as much as 64 percent.[13] In a country where less than 1 percent of hospitals are fully air conditioned, this type of measure saves lives, especially as the annual number of very hot days increases with climate change.[14]

Where air-conditioning systems are commonplace, hospitals and clinics can make other adjustments to account for new climate change-related extremes. For example, health care facilities can ensure they have adequate back-up power and windows that open to allow fresh air inside should those systems fail over extended periods. Other interventions will require the installation of sophisticated equipment. Hospitals operating in wildfire-prone regions may need to install ventilation systems that maintain air quality even under heavy smoke conditions, so that they can continue to treat patients, including firefighters and affected residents.[15] If hospitals and other health care providers fail to prepare, the impact on the surrounding communities will be felt quickly and painfully.

MAKING THE HEALTH CARE SYSTEM SMARTER

In a warming world, a resilient health care system also requires enhanced predictive capacity and effective early-warning systems. Once an oncoming health threat is identified, public health practitioners should have the capacity to mobilize resources to address it, including providing the public with the information it needs to take precautions. As we saw in chapter 5, on data, information technology promises to improve our understanding of climate risks dramatically, and this includes threats to human health. Better surveillance and tracking of the development and spread of diseases allows public health authorities to prepare. For instance, in its aptly named Project Premonition, launched in 2015, tech giant Microsoft partnered with cross-disciplinary researchers across the United States to develop a global system that detects and monitors emerging mosquito-borne diseases before they spread. Early detection has the potential to lower the rate of infection, reduce treatment costs, and cut mortality rates, even as climate change expands mosquitos' habitable range.

How does the system work? Autonomous drones first locate where mosquitos are most active. That, in and of itself, already offers "a revolutionary increase in efficiency," according to one of the researchers.[16] Information provided by the drones is then used to deploy robotic "smart traps" in high-risk zones. By measuring the beat of a mosquito's wings, these traps can distinguish potential disease-carrying specimens from the rest of the known 3,600 mosquito species. The device captures only the disease-carrying variety. Microsoft's cloud computing capabilities then sequence the bugs' DNA to detect pathogens, reducing processing time from thirty days to just twelve hours.

If it is successful, the project could potentially identify *Aedes aegypti* mosquitos that carry the Zika virus and pinpoint where Zika may emerge.[17] "If we can detect these new viruses before they spread," said Ethan Jackson, who is the lead researcher for the initiative, "we may someday prevent outbreaks before they begin."[18] The goal is to set traps around the globe, including in poor countries, where mosquito-borne diseases are killers. This technology could prove valuable not only in searching for Zika, but also for other mosquito-borne diseases such as West Nile, Dengue, and Chikungunya.

Early-warning systems that communicate information about an impending disease outbreak can also save lives by giving both government authorities and the public time to take preventive action. Some researchers even dream about the creation of a global early-warning system for infectious disease.[19] As computer modeling improves, along with the collection of all types of data, deeper investments in early warning hold great promise. But any such investments must consider how climate change could impact the spread of disease.

In 2014, Japan and South Africa embarked on a partnership to create an early-warning system for infectious diseases in South Africa. When implemented, the system will provide the South African government with information based on the

Figure 7   Project Premonition's mosquito "smart trap." Source: Courtesy of Microsoft.

climate-prediction modeling of anticipated outbreaks. This will give the authorities sufficient time to prepare, say, by stockpiling necessary supplies. Focusing on diseases known to be affected by climate change, such as malaria, pneumonia, and diarrheal diseases, the partnership intends to implement the system across South Africa and neighboring countries as well.[20]

Three years after the partnership was launched, an unprecedented outbreak of malaria in South Africa demonstrated how valuable such a system could be. Japanese researchers, in collaboration with a South African think tank, concluded that climate impacts, including higher levels of precipitation and warmer sea-surface temperatures, had altered the geographic spread of malaria.[21] If planners could take advantage of the lag time between climate-related events and the incidence of disease, they could better manage its spread.[22]

Since heat kills more people than any other natural hazard, better heat advisories promise to save lives. Ahmedabad, the Indian city mentioned earlier in this chapter, is expected to suffer from heatwaves of longer duration, greater frequency, and increasing intensity in the coming decades.[23] In 2010, during one particularly ferocious heatwave, the city's seven million people experienced temperatures of nearly 116°F (47°C). Although the India Meteorological Department had forecasted the heatwave two days before it occurred, municipal agencies failed to communicate the message effectively.[24] Over 1,300 people died.

The disaster was Ahmedabad's "no more" moment. Authorities became determined to find a better way to give people more time to take precautions. The city partnered with academic, health, and environmental groups to create an early-warning system and become the first city in South Asia to develop a comprehensive plan to address extreme heat. The city's revised plan, released in 2017, called for improved communication through public messages and social media, text messaging, email, and mobile apps such as WhatsApp. The plan also required new training for health care professionals and the mapping of high-risk areas to show where cooling spaces, drinking water supplies, and other preventive measures should be deployed. Since Ahmedabad adopted the plan, heat-related mortality has declined by 20 percent to 25 percent, according to the Indian Institute for Public Health.[25] The program has spread to multiple cities in India, accompanied by multiday forecasts from the India Meteorological Department.

It's not just the people of Ahmedabad who urgently need early warning and heat protection. Experts estimate that about 30 percent of the world's population is exposed to deadly heat conditions for at least twenty days of the year.[26] Under a business-as-usual scenario, in which emissions continue to rise, it is predicted that by 2100 as much as three-quarters of the world's population will be exposed to the same frequency of deadly heat. In view of this, developing and adopting early-warning systems and action plans for heatwaves is critical. The Global Heat Health Information Network, a group of scientists and policymakers focused on preparing for extreme heat events, took stock of heat action plans across the globe. According to the Network, only about a quarter of the world's countries had heat action plans in place. Over two-thirds of the countries with plans were in Europe, which had its "no more" moment in the deadly heatwave of 2003 that killed more than seventy thousand people.[27]

TEACHING PEOPLE TO TAKE CARE OF THEMSELVES

Cholera has plagued human settlements since ancient times. In 1854, British epidemiologist John Snow discovered that the disease was spread by contaminated water. He traced an outbreak to a single pump near a cesspool in London, where the mother of a baby who had died of cholera had washed the baby's diaper. Caused by the bacterium *Vibrio cholerae* in fecal matter, the disease brings on diarrhea so severe that victims can die of organ failure and shock within hours if it goes untreated. Despite humanity's long experience with the disease, cholera remains a significant public threat and is anticipated to increase due to climate change. Climate-exacerbated flooding spreads the bacteria, while drought can lead the bacteria to accumulate in dangerously high concentrations. The disease affects all age groups, but it is particularly deadly for young children. Concerted public-health outreach can reduce the threat greatly.

Bangladesh is no stranger to cholera. As recently as 1988, 27 percent of flooding-related deaths in rural areas in the country resulted from cholera.[28] Yet thirty years later, when unprecedented floods destroyed close to 700,000 homes, almost no one died from diarrheal diseases, including cholera.[29] What changed in the interim? Bangladesh had launched a proactive campaign to better equip medical facilities and increase public awareness of the need for clean water and basic hygiene and of the use of oral saline to treat diarrhea. Most of the country has access to oral rehydration salts, since even the smallest shops now carry them, and people are well aware of the need to use the salts when diarrhea occurs.[30] Women have been taught to filter water through folded saris, which can effectively remove as much as 99 percent of the pathogens.[31] In the words of Bangladeshi physician Azharul Islam Khan, "Being a country with 160 million people, we are doing fairly well with diarrhea control, I must say."[32]

THE "WICKED PROBLEM" OF CLIMATE CHANGE AND MENTAL HEALTH

The impacts of climate change on mental health are less obvious but no less important. With more people exposed to extreme climate events, psychological challenges multiply. A growing stack of public health studies suggests a strong link between warming temperatures and deteriorating mental health. Evidence now associates heatwaves with increases in population distress, psychiatric hospital admissions, and suicides.[33] For example, researchers have found that warming temperatures are linked to higher suicide rates among Indian farmers and among people in the United States and Mexico.[34]

Climate-related disasters take a toll on peoples' mental health. Almost half of Americans who weathered Hurricane Katrina developed some sort of anxiety or mood disorder.[35] Take, for example, the experience of Denise Thornton. Trapped for days inside New Orleans' Superdome stadium in the wake of Katrina, Thornton and thirty thousand others experienced days crowded into a space that descended into violence, filth, and chaos. The traumatic memories of that time would haunt Thornton long after the storm. "[F]or probably two years after [Katrina] every time I was in a crowd, even at buildings in other cities, I'd have a panic attack," she said. "The sound of a helicopter would throw me into a panic attack . . . And I didn't know why, because I knew I was safe."[36] Climate-related disasters can further lead to post-traumatic stress disorder and depression.[37] Merely contemplating human helplessness before the "wicked problem" of climate change can lead to feelings of despair.[38]

Health care providers, policymakers, and researchers must pay more attention to the nexus between climate change and mental health, especially in cultures that still stigmatize mental illness, including the United States. Governments need to invest in programs like one created in India to reduce the economic and mental stress that comes from losing crops and livelihoods. A $1.3 billion climate-based crop insurance program in that country is aimed at cutting suicides among farmers.[39] Additionally, rapid disaster response can reduce anxiety and stress by shortening the period of time people must wait for aid.[40] For mental health, as for other types of health care, continuity of treatment during extreme weather events is critical.

TRAINING FOR HEALTH PROFESSIONALS

As the Zika crisis demonstrated so vividly, public health officials and practitioners need to understand how to detect, diagnose, and treat conditions they may not have previously encountered as the nature of health threats shifts with a changing climate. Health practitioners who previously never had to worry about a specific vector-borne disease, for example, will need to pay much more attention because climate change may bring those vectors into their communities rapidly.

Unfortunately, training has lagged behind. Medical and public health schools have not yet developed deep expertise on climate change-related impacts. Like architecture schools (see chapter 1), medical and public health schools have lagged in establishing specialized training programs on climate risks for health care professionals. Only in 2012 did a school of public health in the United States establish a formal program on the subject; in this case, it was the Mailman School of Public Health at Columbia University in New York City.

In December 2015, during the United Nations conference that produced the Paris Agreement, over a hundred deans of medical, nursing, and public health schools announced their commitment to including climate change in the health education curriculum. Not long after, the Global Consortium on Climate and Health Education at Columbia University was launched. In 2016, the consortium announced a set of sixteen core competencies that should be incorporated into the training of health professionals, including the ability to recognize and reduce climate-related health effects. The *Lancet*, a prestigious medical journal, identified training the next generation of health professionals "to recognize, respond, facilitate preparedness, and educate others [as] essential to preparing for climate impact on health."[41] The International Federation of Medical Students Association has similarly called for incorporating health impacts into the curricula of every medical school by 2020.

Progress in the field of climate and health needs to be made much faster. In 2015, Margaret Chan, the Director General of the World Health Organization, called climate change the defining issue for public health in the twenty-first century.[42] Yet three years later, the United Nations Environment Programme concluded that efforts to address health impacts fall "well below" what is required to minimize the harm.[43] For any efforts to succeed, health care professionals and public health officials must understand the risks and have access to the tools, training, and facilities to continue to deliver vital care in a changing climate.

PRESCRIPTIONS AND PROVOCATIONS

- Governments should invest in enhanced disease surveillance, predictive capabilities, and early-warning systems to facilitate preparedness efforts and preventive action.
- Hospitals and other medical care facilities should conduct climate-vulnerabiltiy assessments and invest in resilience-enhancing retrofits and upgrades so they can continue to provide care during extreme climate-related events.

- Medical, nursing, and public health schools should promptly incorporate relevant climate change-related material into their core curricula.

- Community leaders, in partnership with academic institutions, emergency managers, and other stakeholders, should develop and implement heat action plans to prepare for extreme heat events in at-risk areas.

[8]

BUFFER GROWING INEQUALITY

On June 15, 2015, days before its official release, someone leaked a much-awaited paper authored by Pope Francis to the Italian press. The Vatican called the leak a "heinous" act and immediately suspended the journalist's access to the Holy See.[1] Much speculation ensued in the press about who had leaked the paper and why. Were conservatives inside the Vatican trying to undermine the papal message? Was it an effort to embarrass the Pope? Would such dirty tricks plague the pontiff throughout the remainder of his tenure?

The leaked document was an encyclical, a letter from the Pope concerning Catholic doctrine. Over the centuries, pontiffs have used encyclicals to highlight priority issues and to shape Church teaching. A message like this one can quickly radiate outward from Rome and make its way into sermons and homilies in thousands of churches around the world. The leaked encyclical was entitled *Laudato Si'* (Be Praised) and subtitled *On Care for Our Common Home*. Colloquially, it was known as Francis's "climate change" encyclical.

In the end, the public quickly forgot the leak, but not the paper's content. Indeed, the encyclical's message resonated widely. "Climate change is a global problem with grave implications: environmental, social, economic, political and for the distribution of goods," wrote

Francis. "Its worst impact will probably be felt by developing countries in the coming decades. Many of the poor live in areas particularly affected by phenomena related to warming . . . They have no other financial activities or resources which can enable them to adapt to climate change or to face natural disasters."[2] In other words, climate change will disproportionately affect the poorest communities and exacerbate global inequality.

With the Paris Agreement just months from adoption, the Pope's encyclical proved timely and shaped the debate across many different quarters. It echoed messages coming from the climate-justice movement, which brings together a diverse set of activists and nongovernmental organizations, many of them nonreligious, who call attention to the social and economic inequities of climate change. The Pope's message also resonated with faiths outside the Catholic Church. Shortly before the encyclical's publication, more than twenty faith leaders convened at the White House to meet with Obama administration officials. They included Sikhs, Muslims, Jews, Catholics, Protestants, and Hindus. Even representatives of the Evangelical community, which is highly influential in US politics but generally skeptical of climate action, attended. The faith leaders at the meeting explained why they favored urgent climate action. As the Pope's encyclical would assert days later, protecting the poor and vulnerable demanded it.

This chapter describes how climate change magnifies existing economic and social inequalities and identifies strategies that can help buffer against this effect. It starts by looking through a wide-angle lens, viewing the nexus of climate change and inequality from a global perspective before homing in on the United States. As is the case with other areas of climate resilience, nobody has all the answers on how to manage climate-exacerbated inequality. Developed and developing countries have much to teach one another on this issue.

THE GEOGRAPHIC LOTTERY

Climate change will touch virtually every country in the world and region of the United States, but it will not affect every place equally. Looking around the globe, we can see hotspots emerging, places that are likely to suffer more than others. Small island states, in particular, face existential risks. Sea-level rise threatens to swallow up much, even the entirety, of some island nations in the Pacific, such as the Marshall Islands and Kiribati. Moreover, droughts, floods, loss of fisheries, and salt water intrusion may force residents to abandon island countries well before these atolls disappear.[3]

According to the most recent research, warming temperatures will hit hardest in countries that fall along a wide arc that sweeps from Brazil in South America, through West and Central Africa, past the Middle East, and then down across South and Southeast Asia, ending in Australia. Countries in this arc are expected to suffer steeper declines in economic growth per person than other parts of the world.[4] This is because they are already starting from higher baseline average temperatures than other countries. Researchers believe that the damage to economic growth will result from a complex combination of factors, including falling crop yields, lower worker productivity, and greater incidence of heat-related health problems.[5]

Superimposed on this "geographic lottery" is the existing pattern of economic inequality across and within countries, which makes matters worse. Indeed, one of the most disturbing aspects of climate change is that it will often affect the places that already happen to be poor and are therefore less able to build resilience most severely. Temperature shocks and their impact on economic activity offer the clearest illustration of this tragic irony. In general, rich countries tend to enjoy cooler average temperatures than do poor ones. The average annual temperature is about 52°F (11°C) in developed countries, while the typical poor country experiences an average temperature of 77°F (25°C).[6] In practice, that means that in places like Germany, France, and the United Kingdom, more warming is expected to cause less damage to the economy than in hotter places. In very cold nations, moderate warming may even provide some economic benefits. Meanwhile, in countries such as India, Indonesia, and Nigeria, high temperatures are already stifling economic growth, and more extreme heat will mean even more damage to economic activity, slower growth, and therefore less capacity to pay for investments in resilience.[7]

This geographic climate inequality shows up in countries that are big enough to span multiple climate zones. Take the continental United States. Experts expect climate change impacts to harm the economies of southern and lower midwestern states the most. Economic damage will result from a combination of coastal damage, wilting crops, higher spending on electricity to cool buildings, and more heat-related health impacts, among other things.[8] Meanwhile, the economies of states in New England and the Northwest may actually benefit from moderately warming temperatures, as crop yields improve and people spend less money heating their homes.

Besides geography, gender inequality is another dimension compounded by climate change. Climate change impacts deprive communities of resources, and social prejudice and cultural norms can impose the bulk of that scarcity on women and girls, with devastating consequences. In one disturbing example, researchers sifted through reams of historic data on the economic and health impacts of typhoons in the Philippines.[9] They found that children did not die at abnormal rates immediately after the storms; there were few "exposure deaths" during the typhoon itself. But, in the calendar year after the storm, more children died than would normally be the case. Most strikingly, girls died at higher rates than boys, especially if they lived in households where they competed with other children, particularly boys, for resources.

The backstory, supported by the researchers' other findings, is not hard to fathom. After a devastating storm, families are left with less income, which translates into less money for health care, food, and education. When families privilege male children over their sisters, scarce resources go to the boys first. The girls are left to absorb the shortfall, and in many cases, that proves fatal. Some 11,300 girls suffer such "economic deaths" in the Philippines each year, the scholars calculate.

A similar narrative has emerged around drought. Scholars studying Indonesian data found that girls born during years of plentiful rainfall tend to experience better health and mental skills as adults when compared to girls born during dry years.[10] Girls born in wet years grew up to be taller, have better health as adults, and complete more years of schooling than those born in low-rainfall years. But the interesting thing is that the researchers couldn't find this variation in men. Boys fare the same regardless of how much it rained the year they were born. As in the Philippines, what is likely happening in Indonesia is that in low-rainfall years, crops suffer, and family income declines. With fewer resources to go around, the boys get served first, and the girls are left with less, which sometimes means too little. These early inequities compound with time, reverberating throughout the women's lives and their communities.

INEQUALITY OF RESILIENCE

In the United States, some groups are much better prepared to cope with climate change impacts than others. That resilience depends on enjoying access to public and private safety nets, the ability to move, and social connectedness. These three things largely depend on households' wealth and socioeconomic status, so access to them reflects existing patterns of economic inequality.

Safety nets are, of course, essential to enable people to deal with all kinds of disruption, including climate change. Traditional safety nets, such as the social security program in the United States, are provided by the government, and anyone can access them. Other safety nets are private, and access to these is unequal. Personal savings, an example of a private safety net, provide a lifeline to pay for food, transportation, medical care, or housing during and after a climate disruption. However, many Americans are poorly prepared in this regard. A 2018 report by the Federal Reserve found that four in ten adults in the United States would not be able to cover an unexpected expense of $400 without borrowing money or selling something.[11] Another government study found that about 40 percent of the 44 million American renter households lacked access to $2,000 for evacuation expenses.[12]

Private insurance offers another private safety net that can prove critical to bouncing back. An insurance payout can enable a business or a family to get back to normal life more quickly. Here, too, the reality turns on wealth. Almost all homeowners in the United States report buying homeowner's insurance, but less than half of people who rent have renter's insurance and could therefore lose their possessions in a natural disaster.[13]

Access to transportation is a second crucial dimension of climate resilience. The ability to get away from wildfires, flooding, heatwaves, and hurricanes can make the difference between life and death. Inequality plays a role here as well. If public transportation shuts down partially or completely, then a chasm opens between those with access to private transportation and those without. As

Hurricane Katrina approached New Orleans, one in five households did not have access to private transportation. Unsurprisingly, those households tended to be poor.[14] Meanwhile, a third of households in the city had access to two, and sometimes even three, private vehicles. It's not hard to imagine which group ended up stranded as the city drowned.

Mobility is especially challenging for the elderly and for people with disabilities. When disasters of the kind that climate change will make more dangerous strike, the elderly and those with disabilities are more likely to find themselves trapped, injured, or dead. One statistic from Hurricane Katrina is tragically telling: Over 70 percent of Katrina-related deaths in New Orleans were among persons aged sixty and over, even though they comprised only 15 percent of the city's population.[15] Many could not evacuate because they lacked transportation.[16] As author David Perry has noted, "Every natural disaster quickly becomes a story about disability."[17]

But mobility in the age of climate change is not just about escaping an imminently approaching storm or wildfire. It's also about having the option to change your permanent address as climate conditions change. Inequality is relevant here as well. Given sea-level rise, wealthier residents will likely choose to abandon waterfronts and move inland to higher ground. In some areas, the value of flood-resilient properties will appreciate until they become unaffordable for many. Meanwhile, the prices of vulnerable coastal properties may plummet, or at least appreciate more slowly than homes uphill. As we saw in chapter 3, on markets, this may already be happening. While prices and rents climb in the more resilient, elevated areas, poorer residents will be stuck in soggy, flood-prone neighborhoods. "Climate gentrification," as some scholars have called it, will exacerbate inequality and leave the poorest people to live in the riskiest places.[18]

Also consider the climate risk borne by places where many of America's poor live already, namely in public-housing complexes. Built in the 1950s, Tidewater Gardens is a public-housing community in Norfolk, Virginia, a city that, as we have seen, is already impacted by sea-level rise. The median annual household income in Tidewater Gardens is $12,000. Not only is the community poor, it is located in one of the most flood-prone areas in one of the most flood-prone cities in the United States. The city has plans to demolish the buildings and redevelop them with flood-protection features. Residents will be temporarily relocated, but much uncertainty remains as to who will get new housing and at what cost.[19]

Tidewater Gardens is hardly an exception. Almost half a million federally subsidized rental and public-housing units (9 percent of the US public-housing units, and 8 percent of the federally subsidized rentals) lie in the combined 100- and 500-year floodplain (the area that has a 1 percent to 0.2 percent chance of flooding in any given year, respectively).[20] Residents who are able to leave these communities and move to safer neighborhoods as climate change advances will likely be better off than those left behind.

Finally, social connectedness is critical for resilience, but in this case, income and wealth are not necessarily the only things that matter. In a famous study of Chicago's deadly 1995 heatwave, sociologist Eric Klinenberg documented the experience of a pair of similar communities.[21] The adjacent neighborhoods, Englewood and Auburn Gresham, were almost entirely African American. Both had similar proportions of elderly residents, and both had high levels of poverty, unemployment, and violence. Yet, Englewood proved to be a death trap during the heatwave, while Auburn Gresham turned out to be safer than many of the affluent neighborhoods in the city, with a mortality rate one-tenth that of Englewood's.

Klinenberg credits the contrast with differences in social connectedness. Yes, Auburn Gresham was poor, but it had busy storefronts, restaurants, and sidewalks that, though modest, kept people in contact with each other. In Auburn Gresham, residents looked out for each other during the heatwave. In contrast, Englewood had lost residents and commerce over the years. Residents were afraid to go out into the unsafe, desolate streets, and neighbors did not know one another. Englewood had become a place in which it was much easier to die alone during a heatwave, especially if one were sick, isolated, or old. Similarly, Canadian public health authorities sifting through data about a deadly 2018 heatwave in Quebec discovered that most of the deaths were of men living alone, especially those suffering from physical or mental-health problems, or experiencing substance abuse.[22] Social isolation kills, and the isolated are often poor or vulnerable.

STRENGTH IN COMMUNITY

While social isolation can prove deadly, social resilience can save lives. Women were less likely than men to die during the Chicago heatwave because women had stronger ties to friends and family. Latinos also did better than other groups despite their low incomes, partly because they lived in crowded apartments and dense neighborhoods where everyone was looked after.[23] Social connectedness makes a big difference during times of crisis.

Fostering social resilience requires promoting bottom-up, organic, local networks of mutual assistance. The experience of neighborhoods like Auburn Gresham in Chicago suggests that safe public spaces and strong civic organizations promote the kind of connectedness that can help save lives during extreme climate events. However, building up trust in neighborhoods where it has weakened (or has never emerged) is no easy matter. There may be no "start-up kit" for social resilience.

In the meantime, governments need better ways to locate the most vulnerable members of communities, especially those who may be isolated, so they can receive assistance first. After the 1995 heatwave, Chicago authorities started collecting data about where the elderly, chronically ill, and otherwise vulnerable live, so city workers could check in on them. The city also developed a system of email, telephone, and text-message alerts to share emergency information. France did something similar after the 2003 heatwave. When a heatwave occurs in France now, local authorities contact the elderly and the homebound to ensure their well-being. Technology can also help emergency personnel decide who should get attention first. Geospiza, a Seattle-based company, relies on artificial intelligence to help city emergency managers find and protect their most vulnerable residents during a disaster. Computers scour and triangulate multiple databases to estimate, for example, which residents may be hearing impaired or use personal-care attendants. The program then notifies authorities, so they can check on these addresses first.

STRONGER BUT NIMBLER SAFETY NETS

In a world of escalating climate risks, the first line of defense for any community will continue to be social protection programs. In the United States, these include traditional programs such as social security, unemployment insurance, nutrition assistance, and health insurance for the poor. These programs will acquire new importance when climate change impacts make displacement and economic shocks more common. It is essential that political leaders protect and reinforce existing safety nets and that everyone who needs these programs can access them.

In parallel, governments should develop a toolkit of temporary measures to deploy in the immediate aftermath of a disaster and give those affected a rapid injection of emergency cash to help them rebuild their lives. After Hurricane Katrina, for instance, financial authorities allowed New Orleans residents to withdraw money from their retirement accounts without incurring early-withdrawal penalties. This change unlocked thousands of dollars in a key moment of need.[24] The federal government also extended tax breaks to companies that hired residents of areas affected by the storm.

Another idea worth considering is insurance-powered safety nets. This involves coupling parametric insurance, a type of insurance that pays out quickly (see chapter 4) with existing systems for delivering assistance to the poor. Government entities in charge of social safety nets could purchase parametric insurance from the private sector and design the policy so that it triggers automatically when an extreme weather event of a certain magnitude strikes a certain location. Government agencies could then use the money from the insurance payout to deliver rapid payments to poor and vulnerable people in affected areas, using existing social-protection systems to deliver the aid.

This approach could prove useful for cash-strapped local and even state governments, especially since federal assistance often arrives with delays. Developing countries could also benefit from insurance-powered safety nets. Kenya's government, for example, has considered combining a parametric insurance policy for drought with its Hunger Safety Net Program, which in ordinary times provides food-insecure families with regular cash transfers of $25 per month through biometric smartcards.[25] If a drought gets bad enough to meet a certain threshold, the insurance policy pays out automatically, and the additional money is delivered to affected families through the Hunger Safety Net Program mechanism.

DECISION-MAKING WITH INEQUALITY IN MIND

In the years following Superstorm Sandy, New York City planners pondered how to make the city more resilient to the next superstorm. A few months after the storm, in the summer of 2013, Mayor Michael Bloomberg announced a $20 billion plan to safeguard the city from the impacts of climate change.[26] Under the motto, "A Stronger, More Resilient New York," the plan called for significant spending on infrastructure to protect vital structures and services. Six months later, Bill de Blasio replaced Bloomberg as mayor. The new mayor, who predicated his political campaign on making New York more affordable and less economically unequal, promptly announced his own plan, OneNYC, under the banner "A Strong and Just City."[27] De Blasio wrapped around Bloomberg's resilience ideas a larger vision of investing in affordable housing, education, social services, and other elements focused on social justice.

The two mayors' approaches highlight a central dilemma in building resilience. Who should benefit from resilience investments? Decision-makers' instinct is often to put resilience dollars where they will avoid the largest economic losses. Because rich communities have higher concentrations of valuable buildings and infrastructure, building resilience in those places will usually generate the highest bang for the buck when measured in terms of economic losses avoided. As one New York City planner told us, "If I wanted to avoid the most losses, I would put every resilience dollar I have to protect lower Manhattan." But of course, this logic exacerbates economic inequality because it leads decision-makers to protect relatively small pockets of concentrated wealth with resources that could be spread to benefit many others. This is both politically and ethically unsustainable. Policymakers must use a different lens in making decisions about resilience.

One approach involves thinking about welfare losses. Disasters don't affect everyone equally. A loss of one dollar is much harder for a poor family to absorb than it is for a rich one; helping a poor family prevent the loss of a dollar has larger benefits in terms of hardship and suffering avoided than does sparing a rich family the loss of that same dollar. Decision-makers need a method for assessing resilience investments that takes this difference into account. The method should encourage policymakers to choose investments that deliver large well-being benefits, not just avoid economic losses.

Some governments are exploring ways to apply this approach. After massive flooding hit the Indian city of Mumbai in 2005, economists at the World Bank looked at two different ways to calculate the costs of the devastation.[28] Simply counting the economic losses—the cost of destroyed property and infrastructure—the experts pegged the total cost at an estimated 35 billion rupees. But that didn't tell the whole story. Using a welfare-loss methodology, they estimated that the "well-being" losses—the economic losses adjusted by how much the loss of a rupee hurts a household's well-being—were almost double the economic losses, about 60 billion rupees (roughly $860 million). To be sure, the methodology is complex, and it still needs testing before it can go mainstream. But there is no question that we need to think beyond the narrow lens of economic loss, or we risk making economic inequality worse as we try to build climate resilience.

NO ELYSIUM

In the popular 2013 movie *Elysium* starring Matt Damon and Jodie Foster, the world's elites flee to a majestic space station orbiting the earth after environmental and social collapse has ravaged the planet. The space station is replete with lush gardens, mansions, and robotic servants. But in real life, enclaves like Elysium don't exist, at least not yet. Companies, including the largest and most powerful, must work together with the communities in which their headquarters, facilities, suppliers, and biggest customers are located to make the whole community resilient.

Dealing with inequality in a time of climate change doesn't just mean protecting the most vulnerable. It also means ensuring that the rich and powerful don't withdraw from society, thinking that private systems will shield them from harm even as the public systems around them fail. Communities need businesses to keep running, just as much as the people who own and lead those businesses need the communities in which they are embedded to be resilient. Companies can "climate proof" their facilities, but they still need their employees to show up to work and their suppliers to deliver. They can't do that if the roads are underwater, if workers' homes have suffered damage, or if employees are too worried about their families' safety to show up for work.

Iconic photos taken shortly after Superstorm Sandy drive the point home. One shows the headquarters of the Goldman Sachs investment bank, in lower Manhattan, shining brightly, virtually every floor lit, amid a forest of darkened skyscrapers.[29] Another shows the same well-lit building surrounded by waist-deep black water, a partially submerged SUV floating in front of it (Figure 8). Goldman had invested in private power generators, and before the

Figure 8 Goldman Sachs headquarters, New York City, October 2012.

storm hit, had stacked 25,000 sandbags around the building. But the problem that still bogged down Goldman's operations was how to get employees to work in a city whose transportation infrastructure had largely stopped working. "We have groups in New York right now talking about that," Goldman Sachs's chief operations officer Gary Cohn told the media at the time, "and there's a bunch of conference calls trying to figure out how to get people where they need to be."[30]

The aftermath of the 2011 Thai floods (see chapter 3) on markets provides an example of how businesses, governments, and local communities can work together to build resilience for everyone. After the disaster destroyed numerous factories owned by large Japanese corporations, the companies, the Japanese government, and the Thai authorities got together to figure out how to strengthen flood resilience. They developed strategies and action plans for flood prevention. The Japanese government funded a flood management plan for the Chao Phraya River and provided technical assistance to Thai personnel to improve water management. Thai authorities backed the plan and worked to implement it. More of this kind of collaboration will be needed in a world impacted by climate change.

Local political leaders have an especially important role to play. They have the authority and legitimacy to bring together key stakeholders and lead an inclusive dialogue. Local leaders can also serve as the driving force in community-wide planning, with the federal government providing support where possible. Crucially, the planning process must include the voices of all the key stakeholders in the community. This way, the process can benefit from local knowledge and ensure that the resulting plans enjoy widespread support.

Climate change will manifest itself unevenly across a world already characterized by great disparities in wealth and income. Poor and vulnerable communities—often hit the hardest by climate change because of their geography and lack of resources to cope with disruption—will suffer most. Deploying a full range of tools and approaches will be critical if we are to prevent climate change from severely weakening America's social fabric and worsening political tensions and polarization.

PRESCRIPTIONS AND PROVOCATIONS

- Governments should bolster traditional social safety nets, as well as temporary safety-net measures that can be deployed in the aftermath of extreme events; federal and state governments should work with the reinsurance industry to pair social-protection programs with parametric insurance policies.

- Mayors, with federal support, should regularly convene local private-sector leaders and other stakeholders to identify climate risks and develop collaborative plans to protect supply chains, workers, and critical infrastructure.

- Governments and the private sector should work together to develop innovative solutions to locate isolated and vulnerable persons and ensure they have access to life-saving information and resources.

- Federal government agencies that provide international development assistance should expand efforts to help other countries in building climate resilience, especially with respect to food and water security, public health, gender equity, and disaster risk management.

- Governments should experiment with methodologies to help ensure that investments in resilience are made based on considerations of welfare impacts, not just economic losses.

[9]

RELOCATE PEOPLE
TO SAFER GROUND

When Super Typhoon Haiyan made landfall in the Philippines, in November 2013, it was the strongest tropical cyclone ever recorded. The Philippines is no stranger to storms, but Haiyan was of a different magnitude—a "beast" in the words of one tropical storm expert.[1] Haiyan hit Tacloban City, a coastal city of 250,000 located southeast of Manila, with full force. The water heaved whole ships on top of houses, wiped out roads and bridges, and killed thousands.

The scope of the disaster overwhelmed the local government and the Philippine military. In Tacloban, things quickly devolved into anarchy. Sewage systems broke down. Shortages of food, water, and medical supplies quickly arose. Looters rampaged, and hundreds of convicts escaped from local prisons. The storm destroyed over a million homes, displacing over four million people, almost half of them children. With shelter available for only a hundred thousand people, evacuation facilities quickly reached capacity. Virtually all the residents of Tacloban were left without a roof over their heads.

As conditions deteriorated in the city, the lack of adequate shelter and basic provisions forced people to move. Two days after the storm had made landfall, people began leaving Tacloban for other parts of the country, at a rate of about five thousand a day. Six days after the storm, that number had grown to ten thousand people per day.[2] Shortly thereafter, President Benigno Aquino III conceded: "The system failed. We had a breakdown in power, a breakdown in communications, a breakdown in practically everything."[3]

Beyond Tacloban, the Philippine government struggled to find lasting solutions to resettle the approximately one million people living in areas at high risk of coastal flooding.[4] The government attempted to establish "no-build zones" to keep people from resettling on dangerous coasts, moving many thousands of people inland to live in bunkhouses while they waited for more permanent housing. But the authorities ran into numerous obstacles. Figuring out who needed to be resettled and where they should move proved to be complicated. Confusion over muddled roles and responsibilities among national and local government entities hampered the effort. Mayors shrank away from enforcing the no-build zones and from barring people from returning when few resettlement sites were available.

Families that returned to the no-build zones found themselves in makeshift structures made of salvaged debris, which often put them in even greater peril. "NGOs keep telling us that this is a no-build zone," an elderly woman living by the water in Tacloban was quoted as saying. "But I have no other choice. I'm just waiting to be relocated, but I don't know when."[5] Three years after the storm, the country had met only about 1 percent of its target goal of finding homes for displaced households.[6]

Whether the world is prepared for it or not, climate change will drive large-scale migration. The impacts of climate change—both slow-onset changes, such as sea-level rise and drought, and sudden-onset events, such as extreme storms and wildfires—push people from their homes. Indeed, the US National Intelligence Council has concluded that the "net effects of climate change on the patterns of global human movement and statelessness could be dramatic, perhaps unprecedented."[7] In 2018, the World Bank projected that by 2050, climate change could create over 140 million internally displaced people in developing regions across the globe.[8] In the United States, sea-level rise could force millions of people to relocate.[9]

Managed well, migration can yield enormous benefits, offering greater opportunities for those who relocate and injecting new talent and energy into receiver communities. But climate change threatens to unleash what the National Academy of Sciences in the United States terms "disruptive migration"—that is, sudden migration that could strain social, economic, and political stability.[10] The task ahead in the face of climate change is to encourage managed, gradual migration that minimizes disruption, moves people out of harm's way, and turns displacement into economic opportunity.

The strategies and tools to make this possible exist. Technology and data analysis can enable us to figure out the areas that are unsalvageable and unsuitable for repopulation. Governments can provide incentives for homeowners to leave, identify areas suitable for relocation, and devote funds to support people on the move. Incentives provided to receiver communities can be in the form of tax breaks for those that employ migrants and community grants that establish systems and institutions to accommodate and help integrate newcomers. Funding new infrastructure can help handle the needs of the growing population. The private sector can scale up operations in the receiver communities to create economic opportunities for newly arriving migrants. Revisiting laws and international arrangements to facilitate migration in response to climate change could reduce pressure on international borders. Putting strategies in place in advance of the crisis is imperative, and leaders must develop political strategies to talk about what will surely be a very controversial issue.

FOR SOME, DISPLACEMENT WILL
BE INEVITABLE

Displacing people from their communities is traumatic, even in the best of circumstances. This is why it typically makes the most sense to help people stay put. Keeping communities intact enables economies to grow, protects the tax base, keeps kids in school, maintains social cohesion, and avoids the emotional and physical toll of displacement. It also avoids putting an unexpected and unplanned burden on the receiver communities. Many of the policies outlined in this book can help keep residents safe at home as climate change advances. But climate change also means that some people will inevitably have to move. Sea-level rise is the most obvious illustration. According to the IPCC, "migration is the only option" when it comes to sea-level rise—the ground will literally disappear from underneath our feet.[11]

The delta country of Bangladesh has become the poster child for this inevitability. It's one of the most densely populated countries in the world, cramming more than 160 million people into an area the size of the US state of Iowa. With its low-lying elevation and lattice of crisscrossing rivers, the country faces chronic risk of flooding. According to some estimates, if sea levels rise by 3.3 feet (1 m), they could displace twenty-five million or more Bangladeshis.[12] And before that occurs, salt water intrusion can spoil fresh water supplies and degrade the quality of farmland.

Many Bangladeshis will have to find a new place to live. But where? India surrounds the country on three sides; the two countries share a 2,500-mile-long border (4,000 km) partially composed of impassable mountains. Along the rest, India has constructed an eight-foot-high barbed-wire fence that is electrified in some areas.

India appears serious about keeping the Bangladeshis out. In 2010, Human Rights Watch reported that Indian border security forces had killed over nine hundred Bangladeshis in the previous decade.[13]

Bangladesh is certainly not unique in facing this challenge. According to Mathew Hauer, a professor of geography at the University of Georgia, Americans will also have to move in large numbers because of sea-level rise.[14] Forty percent of the US population already lives in densely populated coastal areas.[15] In the coming years, millions of people may leave places such as Miami for Austin, Orlando, Atlanta, and Houston. Receiving cities could each experience a net population increase of more than 250,000 people. As population increases along with sea-level rise, by 2100 as many as thirteen million people in the United States may find themselves moving out of their homes that are flooded are or at risk of flooding.[16]

ENCOURAGING PEOPLE TO MOVE

Ellicott City, Maryland was officially founded in 1772 between two rivers, and it has suffered severe flooding ever since. The city has had to rebuild itself over and over again, using its past experiences to guide the recovery efforts. But flooding has worsened considerably since 2010. Two so-called 1-in-1,000-year floods swept through the town in 2016 and 2018 following torrential rains. The waters ripped through the historic Main Street, damaging buildings and causing tens of millions of dollars in losses. After the 2018 floods, Ellicott City business owners again considered their options. Some wanted to stay where they were and dug in; others wondered whether it was worth starting all over again in the same spot.

In August 2018, county official Allan Kittleman announced a $50 million plan to reduce flooding. It called for the demolition of particularly at-risk historic buildings at one end of Main Street— some dating from the 1800s. The plan was met with mixed reactions, and in November of that year, Kittleman lost his bid for re-election. Within weeks, his successor made clear his intent to reconsider the proposed demolition. Ellicott City serves as a vivid reminder that the decision to abandon structures and alter familiar places does not come easily. People develop emotional ties to the land and to their communities. As the Fourth National Climate Assessment points out, people feel a strong connection to the land they call home and are inclined to remain there, even when their property and livelihoods are threatened.[17]

One way to ensure that climate migration is less disruptive and more likely to lead to a better future for those who are displaced is to encourage them to move well before climate impacts force them to do so. For the most part, the United States has focused its efforts on encouraging people to move out of harm's way through volun- tary home buyout programs. Under these federally and locally run programs, individual homeowners may sell their homes to the gov- ernment and then move to a safer location of their choosing. These programs can thus help to avoid significant losses. For example, a study of twelve communities in Iowa that were damaged by flooding in 2008 concluded that the government's acquisition of over 1,500 at-risk properties in the prior two decades had avoided losses of close to $100 million, which comes out to a cost-benefit ratio of more than two to one.[18]

But the buyout programs have struggled to enable people to relocate on a significant scale. Sometimes people simply refuse to take advantage of the program. For instance, in 2012, the state of Maryland offered buyouts to residents of Smith Island. The tiny island sits on the Chesapeake Bay, just a few feet above sea level. Over the last century, the sea has claimed over 3,300 acres (1,400 hectares) of the island, and with rising sea levels, the remainder could disappear as early as 2025. Yet when the Maryland authorities made their buyout offer, most of the residents said, "No thanks." Instead, they have continued their attempts to slow coastal erosion by constructing new jetties, an effort counterproductively backed by a different source of government funding.[19]

Sometimes, buyout programs struggle because they fail to gain support from a critical mass of residents. In 2013, for example, New York Governor Andrew Cuomo proposed spending $400 million to buy up homes in damaged areas to prevent another round of destruction like that experienced during Superstorm Sandy the year before. The plan envisioned turning properties into bird sanctuaries, parklands, and coastal dunes to buffer against sea-level rise, extreme storms, and coastal flooding. But state and local leaders failed to attract enough interest to create the wide, contiguous swaths of public land needed to protect inland property. Because too few homeowners took advantage of the program, some communities were left with a scattering of condemned properties mixed with occupied homes, and the government was forced to maintain infrastructure to service the remaining properties. Other communities decided not to participate at all because they feared losing the tax revenue generated by the properties once the government purchased them.

As floods and wildfires continue to spread destruction, our innate optimism will eventually give way to realism, and when that happens, the demand for home buyouts will quickly exceed available funds. Indeed, it is already apparent that the money currently dedicated to buyout programs will not be nearly enough. Since 1985, the Harris County Flood Control District, the main agency engaged in buyouts for flood-prone properties in the Houston area, has spent over $340 million to purchase close to 3,100 properties. Before Hurricane Harvey struck, the county had a list of 3,300 homes designated as priorities for the buyout program. Yet, according to the *Texas Tribune*, it would take three decades for the agency to buy all the homes on the list at the pace the county was moving.[20]

Buyouts are a useful approach, but they are a partial solution at best. In some places, it will be necessary to relocate entire communities. As sea-level rise continues its relentless advance, governments will have to search for new ways to help citizens find more hospitable places to live before they are forced to do so. This is already happening. The Pacific archipelago of Kiribati offers perhaps the most iconic example of government-planned, climate change-related relocation. With a population of a hundred thousand, Kiribati consists of thirty-two atolls that are just a few feet above sea level.

Figure 9 Staten Island, New York, home damaged by Superstorm Sandy with signs seeking a buyout, October 29, 2013. Source: © AP Photo/Seth Wenig.

As high tides surge further inland, Kiribati's fresh water supply is at risk of salt water intrusion. Eventually, if the seas rise high enough, the island may be completely submerged. In 2014, the president of Kiribati made headlines when he announced that the country had purchased 5,500 acres (2,200 hectares) of land in Fiji as a potential resettlement site. Ironically, Fiji faces its own sea-level rise challenges and plans to relocate dozens of villages to higher ground.

In the United States, the federal government funded two experiments to facilitate the migration of communities en masse due to climate change-related impacts. One case involves the Isle de Jean Charles, an island off the coast of Louisiana that is disappearing into the ocean. When the project comes to fruition, and if the project designers have their way, the island's residents will relocate to one of the most climate-resilient developments in the country. New homes will be elevated against flooding and extensive green infrastructure will absorb flood waters and storm impacts. The other case involves the tiny fishing village of Newtok, which is one of several Alaskan tribal villages that suffers from severe coastal erosion. The village has no time to lose. It is slipping into the Bering Sea at the rate of about 70 feet (21 m) per year.

The exorbitant costs of relocating communities in these experiments are becoming clear. In 2016, the federal government awarded a $48 million grant to relocate the eighty residents of Louisiana's Isle de Jean Charles to the mainland, which works out to a remarkable $600,000 per resident. In 2018, Congress provided the 350-some villagers of Newtok with $15 million. The Army Corps of Engineers has estimated that it would cost hundreds of millions of dollars to relocate another native Alaskan village, Kivalina (see chapter 2), which unsuccessfully sued for climate damages. When these costs were presented to government officials at a 2014 scenario-planning meeting in Anchorage, the eye-popping estimate caused one Alaska official to whisper in the ear of his federal counterpart, "What do you plan on telling Miami if you pay that for just one village?"

While it is too early to tell how these relocations will turn out, the staggering costs involved will likely put the brakes on any widespread adoption and scaling up of the approach. American and other communities will need to engage their political leaders in a serious dialogue about how to share the financial burden of relocation more widely and how to relocate communities at a lower cost.

DISCOURAGING PEOPLE FROM STAYING

What can be done in the meantime, aside from voluntary buyouts? At a minimum, governments should do more to discourage development in risky areas such as floodplains and more clearly communicate to residents that if they choose to stay, they do so at their own risk. For example, the federal government could refuse to provide flood insurance to new developments in floodplains, as discussed in chapter 3. The government could also deny federally backed mortgages to purchasers of newly built housing in at-risk areas.

For communities that successfully implement buyout programs, the government could make additional funds available to attract even more homeowners into the program. It could also provide targeted financial assistance to towns that are losing residents to temporarily ease the shock of decreased tax revenues. Moreover, a pledge by the federal government to contribute more to a community's disaster-recovery costs in exchange for a successful buyout program could encourage a skeptical community to move forward. Similarly, providing monetary incentives to communities that create actionable managed-retreat plans could encourage greater participation in home buyout programs and spur people to action. Piloting multiple methods now will inform future efforts and increase the chances of success.[21]

Accomplishing all of this will no doubt be challenging. Paul Fraim, Norfolk's mayor from 2004 to 2016, claims to be the first mayor in the nation to admit, while in office, that portions of his town will have to be abandoned. On a news program, he conceded that as flooding increases, parts of Norfolk may have to be declared retreat zones.[22] Norfolk has not yet formally embraced the controversial practice of managed retreat. As other communities have, Norfolk has partnered with FEMA to move people out of harm's way voluntarily, but those experiences have met with mixed results. FEMA and Norfolk authorities worked with homeowners to purchase houses in an historic neighborhood named Riverview, which has experienced frequent flooding, razing century-old houses and turning the land underneath back into marshlands. Some homeowners willingly sold, but others decided to stay. As one holdout explained, "Until nature tears it down or the city makes us move, we'll be here."[23]

California's initial attempts to develop a managed-retreat strategy vividly illustrate how quickly things can turn ugly in a town when property values are at stake. The California Coastal Commission recognized managed retreat as a strategy for communities dealing with coastal erosion in draft policy guidance issued in July 2017.[24] But when the small but affluent sea-side town of Del Mar in southern California raised the idea of managed retreat to address its erosion challenges, some wealthy residents voiced objections to the very concept, fearing the value of their multi-million-dollar homes would suffer. Under intense pressure, Del Mar's City Council ultimately rejected the idea.[25] The town later passed an ordinance that promised to fight the Coastal Commission's efforts to impose managed retreat.[26]

STRENGTHENING
RECEIVER COMMUNITIES

At the opening of Matthew Glass's 2009 novel, *Ultimatum*, a man named Joe Benton has just been elected the forty-eighth president of the United States.[27] The year is 2032. In the two years running up to the election, extreme flooding on the Gulf Coast had made relocation a major campaign issue. Benton was quick to incorporate the issue into his campaign strategy: "No fancy sales pitch, no window dressing, just the simple things in simple language. Health. Education. Relocation. Jobs. A four-part package, New Foundation." In his acceptance speech, Benton promised, "Each one of our fellow Americans who must uproot themselves will find a better life among us—not worse. They will find the warm hand of friendship—not the cold shoulder of hostility. Communities that welcome them— not shun them."

President Benton may be a product of fiction, but his concerns about the economic and social tensions that relocation will generate are founded in reality. As climate impacts force people from their homes, the challenge will be to manage migration in a way that avoids inflaming social tensions and conflict. Climate migration will put pressure on local economies and on employment in the receiver communities. It could affect the ethnic or religious composition of those communities, creating or exacerbating social frictions.[28] And climate migration can place excessive burdens on government and local resources, especially if the receiver communities are already under economic stress.[29] The stresses of integrating displaced populations will affect both poor countries like Bangladesh and affluent ones like the United States.

Hurricane Katrina gave us a taste of the economic pressures that climate migrants will place on receiver communities. When the hurricane blasted through the US Gulf Coast in 2005, it resulted in one of the largest displacements of Americans in the history of the country—over one million people. Two weeks after the storm, evacuees had made it to thirty-four states plus the District of Columbia, with many thousands landing in the neighboring states of Texas and Arkansas.[30] An estimated quarter-million people found their way to Houston; some had traveled there in a five-hundred-bus caravan from flooded New Orleans. About 150,000 of the evacuees were still there a year later, increasing the total population in the Houston metropolitan area by almost 4 percent.[31]

The economic impacts of Katrina's migrants on Houston were modest but real.[32] One study found that the influx of Katrina evacuees was associated with a 1.8 percent drop in wages for local residents and a 0.5 percent decline in Houstonians' probability of being employed.[33] Although Houston's large economy absorbed these effects, there is no telling how future receiver communities might react. If the receiver community is not as prosperous as Houston, is already resource-stressed, or experiences a greater influx of internally displaced people, the impacts could be much more severe. If recent history is any guide, politicians and the media could take advantage of the economic disruption to blow these effects out of proportion, creating a toxic environment for the newcomers. Social tensions and even violence toward migrants could increase as the situation deteriorates.

Typhoon Haiyan illustrated another dimension of the problem, the fact that climate migrants may not be able to make a living in their new community. In Haiyan's aftermath, the Philippine government made a plan to move those who had lost their homes to areas further inland, away from the coast. The plan, conceived in the chaos of the storm's aftermath, met with resistance from some of the citizens it was intended to benefit. The relocation sites were far from the coast, forcing the fishermen who had lost their homes to commute long distances. In addition, there was no obvious way to earn a living. As one official noted, the displaced kept asking, "What will we do in the mountains?" Lack of access to livelihoods drove many members of relocated families to return to where they had come from along the coast. Moreover, relocation decisions had to be made with distressingly little notice, and families had to commit "sight unseen." Ultimately, these challenges undermined the program.[34]

The key, then, will be to preemptively strengthen receiver communities that are likely to absorb people displaced from climate-damaged areas in the future, be they flooded coastal regions or burned-out communities in the wildland-urban interface. It's not just coastal communities that will have to adapt; it's also the inland communities that take in people displaced by eroding coasts.[35] If communities fail to plan now for migration inland, migrants could place significant burdens on receiver communities by century's end, including in places like Phoenix, Las Vegas, Atlanta, and southern California, which already struggle to manage resource scarcity and growth challenges.

Bangladesh is already thinking ahead. To ease the transition of climate migrants inside its own borders, the country embarked on a collaborative effort with the German government to improve living conditions for displaced people who are moving to slums. The project is centered on the metropolitan areas of Khulna and Rajshahi, the third and fourth most populous cities in Bangladesh, respectively. The program established migrant training programs for phone repairers and motorbike mechanics and worked with municipal leaders to improve local infrastructure in slums. Some priority projects included building roads, installing public toilets, and improving access to water and electricity.[36]

Bangladesh is also thinking on a larger scale. By sharing the lessons learned about relocation, the governments and communities on the front lines of climate change can assist cities that are facing similar challenges before they encounter them. For example, Saleemul Huq, director of the Bangladesh-based International Centre for Climate Change and Development, has identified a dozen inland cities that, with support from the government and the private sector, could absorb a million migrants each.[37] He has said that Bangladesh could show the world that planned migration can prove a successful and transformative response in the face of climate change. When preparing for climate change, "the rich can learn much from the poor on how to be resilient," he says.[38] Meanwhile, Caroline Lewis, founder of a nonprofit organization focused on educating the public about climate change, stresses that lessons from successful managed retreat need to be shared. If managed retreat from the coast can be done successfully, "then we would have accomplished a great deal that the whole world could learn from," says Lewis.[39]

Policies that ease the resettlement process and enable displaced people to build a new life in receiver communities will be vital. To this end, the International Monetary Fund has called for strengthening labor markets to assist migrants in finding work, increasing access to affordable education and training, making it easier for foreign qualifications to be recognized, and reducing the barriers to setting up new businesses.[40] Ensuring that climate migrants have seamless access to financial services when they move from one community to another will also help them resettle more quickly.

Managing climate-induced migration can spare a great deal of suffering. As one 2011 study observed, "[A]s many people could move *into* areas of environmental risk as migrate from them" in the coming years.[41] By identifying and working with potential receiver communities, a nationwide managed-retreat plan could reduce the risk that people move to even higher-risk areas than the places they leave behind. The strategy would need to direct investments toward the regions most likely to receive migrants, identify required social services, and steer migrants toward safer areas. Finally, a national strategy could lay out measures that help displaced communities integrate into their host communities. Despite the urgency, no such strategy currently exists in the United States, or for that matter pretty much anywhere else in the world.[42]

REVISITING INTERNATIONAL MIGRATION RULES

Countries have developed principles to govern some aspects of cross-border migration. But to date, no international mechanisms have emerged to determine the status of climate refugees who cross international borders. Under international refugee law, individuals fleeing persecution can obtain refugee status and thus legal protection. People who seek to cross an international border because the land beneath them has eroded away or because their livelihoods have disappeared after salt water contaminated their farmlands are afforded no similar protection. This omission may create a global class of stateless persons as the impacts of climate change worsen.

In 2018, the United Nations sought to tackle the problem of climate refugees and proposed a Global Compact for Safe, Orderly and Regular Migration. The United Nations General Assembly endorsed the compact, with 152 countries voting in favor and five, including the United States, voting against. As the compact is not yet an international treaty, it remains nonbinding. Without international consensus, the limitations of international refugee law will only come under increasing pressure. For example, an examination of asylum applications in Europe between 2004 and 2014 found that when temperatures in migrant-sending countries deviated from the optimal temperature for agriculture, the number of asylum applications increased by about 351,000 per year.[43] Drawing from these findings, the researchers predicted that asylum applications will likely increase between 28 percent and 188 percent by 2100, as global temperatures rise.

New Zealand has attracted international attention for its attempts to grapple with this issue. In 2015, New Zealand deported a Kiribati native, Iaonae Teitota, after he lost his case seeking refugee status. He applied for asylum on the grounds that sea-level rise threatened his home in Kiribati, making it unsafe for him, his wife, and his three children to return. The New Zealand Supreme Court rejected his bid, ruling, "In relation to the refugee convention, while Kiribati undoubtedly faces challenges, Mr. Teiota does not, if returned, face 'serious harm.'"[44] In 2017, with a new prime minister in place, New Zealand became the first country in the world to announce plans to create special refugee visas for those forced to migrate as a result of climate change. But the idea didn't gain traction, and by 2018 it was all but dead politically.

DEVELOP A POLITICAL STRATEGY

Of all the hard lessons in this book, managing climate migration may be the hardest. Officials in the Obama White House discussed creating a federal policy on managed retreat, but they ultimately dropped the issue for more politically palatable initiatives. We are under no illusion that endangered communities will welcome their own mass relocation, or that receiver communities will find it easy to embrace newcomers. We are also aware that many politicians will avoid talking about and dealing with this issue at all costs, out of fear of being seen to be "giving up" on their communities. And yet, as with all the other resilience lessons in this book, to ignore its inevitability is to leave millions of people woefully unprepared. Our political leaders will need to begin a national conversation on this sensitive topic, and the sooner, the better. The earlier we start, the easier, less costly, and less traumatic the process will be.

Such a political strategy will need two things. First, it must position "preventive action" as an opportunity worth seizing, not as a threat to be avoided. In Matthew Glass's *Ultimatum*, cited earlier in the chapter, fictional US President Benton plans to inject significant government funds into the receiver communities and to build new infrastructure, while letting the infrastructure in coastal areas degrade on its own. "Add to that universal health care, adequate education and job stimulation, and put the whole thing in a reasonable time frame, so you can phase relocation, and that actually gives you growth," an adviser tells the president. This approach gives Benton a politically acceptable message, one of hope and self-determination instead of despair and helplessness. A successful message in real life may not look so different. The state of Louisiana, one of the most flood-prone places in the United States, issued a massive report in 2019 that aims to prepare receiver communities to absorb climate migrants and spur economic growth in the process.

Second, the strategy will need to consider what institutional mechanisms must be created to deal with the relocation challenge.

In *Ultimatum*, the federal government creates new institutions to coordinate the various government departments: a National Relocation Commission headed by a cabinet-level official, as well as a National Relocation Council modeled on the National Security Council. Perhaps that's what political leaders will need down the line. But in the immediate future, existing federal agencies can make a lot of progress within their existing mandates. The Departments of Homeland Security, Housing and Urban Development, and Interior will have to start thinking seriously about a future in which people will need to move. Similar measures will be needed in other countries, and we must all learn from one another. The earlier we start this conversation, the easier it will be to make the truly hard decisions when the moment arrives.

PRESCRIPTIONS AND PROVOCATIONS

- The federal government should develop a national strategy to provide coordinated government services for transit and re-settlement of displaced persons, which includes mechanisms to transfer health insurance, job training and placement assistance, cross-jurisdiction recognition of professional licenses, tax relief to cover relocation-related expenses, and seamless access to financial services.
- The federal government should plan for the creation of a National Relocation Commission to oversee policy development for relocation, coordinate relevant agency activities, and implement the national strategy.
- Governments, in partnership with academic institutions and other stakeholders, should identify potential receiver communities and develop plans to finance and make strategic

investments in infrastructure and social services in those communities.

- The federal government should provide incentives to communities to establish "no-build" zones to preclude development in areas facing high risk of climate impacts.
- The international community should reach agreement on the status of permanently displaced persons whose state of origin is uninhabitable or severely threatened by climate change.
- Business leaders should consider ways to scale up operations in areas likely to receive persons displaced by climate change and to assist their own workers in making a successful transition.

[10]

RECONCEIVE
NATIONAL SECURITY

In March 2015, in New Delhi, India, an American think tank gathered two dozen experts from around the world to play games. The experts included renowned scientists, retired military personnel, diplomats, and national security professionals. CNA, a Washington based-research institute, had designed the games with a basic goal: to explore how the behavior of governments and their leaders might change as the severity of climate impacts increases and global average temperatures continue to rise.[1] CNA assigned each of the experts the role of a senior decision-maker acting on behalf of a particular nation or region. The players would make decisions for their countries on a wide range of policy matters, including whether and how to limit carbon emissions. To ensure that the game would capture climate impacts that will unfold over a long time horizon, it spanned the course of a century; each turn represented a decade.

As the games advanced and the impacts grew worse, the players' behavior began to change. Somewhere around mid-century, the games crossed a tipping point. Players started the games looking for cooperative solutions but later grew "selfish, more insular, and more willing to preserve their status quo."[2] By the time they got to round six, players began to exhibit a "global fatigue with failed states and migrants." The effects of large-scale migration influenced countries' foreign policies and shaped how they viewed their national culture. Concerns about meeting food, energy, and water requirements dominated the players' thinking. As resources grew scarce, some players placed trade restrictions on food and fuel exports to ensure there would be adequate domestic supplies.

The technologists among them began calling for geoengineering, efforts to manipulate the planet's systems to reverse or slow rising global temperatures. Initially, some players wanted nothing to do with a discussion about geoengineering because they felt it carried unknown risks and potentially negative consequences. But as climate change grew worse, some players' objections to geoengineering weakened, and as they were growing weary of sending aid to other nations, they became more amenable to technological solutions. Resource competition driven by climate change led to deepening nationalism and, at least within developing countries, a growing focus on domestic policies rather than international cooperation. By the end of the games, the fictional world order was hurtling toward conflict.

No one who started the game expected to see a near-total breakdown in cooperation as temperatures rose. At least, that was certainly not where the players had intended to end up. And yet, as conditions grew worse, the players resorted to increasingly desperate and unilateral measures, just at a time when cooperation was needed most. The behaviors that surfaced during the New Delhi games—the inward turn, the growing hostility to migrants, and an increasing tolerance for dangerous, last-ditch technological solutions—may well intensify in real life if we do nothing. If policymakers don't pay attention, nations may very well end up where the simulation participants did. In addition to cutting emissions, nations need to avoid falling into self-destructive thinking and focus on how to strengthen resilience across the globe to reduce the disruption and resource competition that threatened to tear the simulated world apart.

To some, the connection between climate change and national security may appear tenuous. What the military understands best is strategy, tactics, and operations aimed at erecting defenses against, or achieving supremacy over, a hostile country or group. Dealing with a faceless, stateless, abstract enemy like climate change doesn't come naturally to the armed forces. One US soldier put the challenge bluntly when speaking to a policy official who had just finished expounding on the threats of climate change: "I understand what you are saying, ma'am, but my job is to defend the hill or to take the hill. What do you want me to shoot?"

To respond successfully to the challenge climate change poses to global stability, the national security establishment in the United States and other countries will need to think beyond weapons deployment and other traditional national security concepts such as the balance of power, national sovereignty, and territorial integrity. Interstate conflict will naturally remain its central preoccupation. But climate change threatens to drastically reshape the security environment, reshuffle geopolitics, and upend some traditional assumptions about what it means to prepare for and win wars. This chapter focuses on climate change as a global disrupter of military might and social order.

THE GREAT DISRUPTER

For years, alarm bells have been ringing throughout the US national security community regarding climate change. Recent secretaries of defense, including President Trump's first secretary of defense,

James Mattis, have concluded that climate change constitutes a national security risk. National security agencies have produced stacks of reports identifying it as such: the Departments of Defense, Homeland Security, and State have each repeatedly enumerated the threats climate change poses to the United States' security in strategic digests. In 2016, the US National Intelligence Council, the de facto think tank of the US intelligence agencies, made public a report concluding that "climate change and its resulting effects are likely to pose wide-ranging national security challenges for the United States and other countries over the next 20 years."[3]

Concern has also mounted in other parts of the globe. An estimated two-thirds of nations around the world have explicitly identified climate change as a national security concern.[4] In 2007, the United Kingdom chaired the first debate at the United Nations Security Council on the impact of climate change on peace and security. In 2011, another debate took place at which Germany was the chair. Italy, Sweden, and the Netherlands have all promoted greater focus on climate security as part of their terms on the Security Council. In 2018, the European Union's Foreign Affairs Council voiced concerns about the implications of climate change for international security and stability, as did the French Ministry of Defense. The Netherlands chief of defense, General Tom Middendorp, has asserted that there is "no security without climate security."[5] On the other side of the world, New Zealand has issued public-policy statements warning that climate change impacts could reduce military readiness; Australia has acknowledged the climate threat to its national security in strategic documents; and Japan has convened global security experts to study emerging climate-related security risks in the Asia-Pacific.

Why have so many governments sounded the same alarm? They expect that climate change will act as the great disrupter, scrambling many traditional notions of national and international security and forcing a reappraisal of how classical questions of war and peace could be reshaped by a rapidly warming world.

CRUMBLING FROM WITHIN

Pakistan, a nation of two hundred million people, ranks as one of the ten nations most vulnerable to climate change in the world.[6] It's not hard to understand why Pakistan earned the ranking. The country regularly contends with heavy monsoons, which lead to severe flooding. Those monsoons have begun to migrate to other parts of the country as a result of climate change.[7] The country also contends with scorching heat. In 2018, Nawabshah, a city of over one million people in southern Pakistan, experienced the highest temperature ever reliably recorded in the month of April anywhere in the world, 122°F (50°C). Three years earlier, a heatwave killed 1,200 people in the city of Karachi, the economic hub of the nation. Rising seas could completely submerge that delta city in the next three to five decades.[8]

Recurring droughts and salt water intrusion into critical fresh water supplies pose severe challenges for the country. Increasingly saline waters have seeped into the Indus River delta, harming fish spawning, agriculture, and the health of local mangrove forests. With rapid population growth, per capita water availability has declined dramatically and is projected to fall even further in the coming decades as a result of climate change, affecting lives, livelihoods, and social stability.[9] Meanwhile, India, to cope with its own climate-exacerbated water woes, threatens to build three new dams that could cut Pakistan's access to river flows.[10]

Whether Pakistan can navigate its climate challenges successfully matters immensely to international security. Since the attacks of September 11, 2001, the United States has attempted to enlist Pakistan as a key ally to fight the spread of terrorism. Yet, the real threat to Pakistan may not be terrorism but climate change, which could act as a great disrupter to the international community's efforts to contain terrorism and reduce tensions between the nuclear-armed neighbors, Pakistan and India.[11]

Devastating floods in July 2010 demonstrated how this disruption could play out. Almost 12 feet (3.7 m) of rain fell in a single week in a Pakistani province that historically averaged slightly more than 3 feet (1 m) of rain per year. Pakistan's meteorological department concluded that the severity of the flood could be partially attributed to climate change.[12] The flooding displaced over fourteen million people, left an estimated eleven million homeless, and damaged thousands of miles of railway. Despite the crushing losses, including the deaths of over 1,700 people, the government's response to the crisis was slow and disorganized.[13] When the president, Asif Ali Zardari, left for a ten-day trip to Europe following the flood, many saw it as further evidence of the government's indifference to the humanitarian crisis.[14]

With little government assistance in sight, groups linked to radical Islamist groups, including the Taliban, moved in to provide meals, water, and medical care. Their presence in the disaster areas sent a powerful message: We are here to take care of you, even if the government is not.[15] The move allowed extremists to fill the power vacuum left by Pakistani authorities.[16] The threat posed by the groups' quick response was not lost on President Zardari, who cautioned that "negative forces" might exploit the country's desperate situation. Zardari warned that extremists would "take babies who become orphans and then put them in their own camps, train them as the terrorists of tomorrow."[17] When a government fails to protect the basic security of its people in the wake of climate-exacerbated disasters, the resulting vacuum can open spaces for violent groups that victimize individuals, inflame nationalism, undermine alliances, and export extremism. State authority begins to crumble from within.

BLOWS TO THE POWER CENTERS

In the fall of 2013, a local official from Norfolk, Virginia, visited the White House to plead for federal help to tackle a growing problem, namely persistent flooding. As we saw in the introduction, Norfolk sits along the southeastern coast of Virginia, in a region known as Hampton Roads, which is home to close to twenty-nine military facilities. The crown jewel, Naval Station Norfolk, is the largest naval complex in the world (Figure 10). Many of the military facilities clustered in Hampton Roads lie at, or just slightly above, sea level.

Figure 10 Naval Station, Norfolk. Source: Courtesy of the United States Navy.

Along that section of the Atlantic coast, sea-level rise occurs three to four times faster than the global average. The Norfolk city official explained to White House officials that floods had begun to gnaw at the edges of the city bit by bit, year after year.

Over the past several decades, the frequency of so-called sunny-day flooding, tidal flooding that inundates low-lying areas in the city, has increased in Norfolk from one or two days per year to ten days per year. Ninety percent of local military personnel live off-base, and routine flooding and tidal inundation regularly affect their ability to get to and from work. At the White House, the Norfolk official shared his worries about the present and future impact of these floods on the local residents. He also shared a deeper worry—that flooding might jeopardize the ability of this center of military power to protect the United States and that extreme weather events might hinder military readiness.

Hurricane Michael provided a dramatic example of how extreme weather events can damage military assets when it swept across northern Florida in 2018. In its path lay Tyndall Air Force Base, home to a fleet of F-22 stealth fighter planes belonging to the US Air Force and costing about $330 million each. As Michael charged toward Florida, military personnel evacuated, and the Air Force began moving the F-22s out of harm's way. But not every plane made it out in time. Seventeen of these fighter jets were left behind, as they had been grounded for maintenance and had to ride out the storm without much more than an aircraft hangar as cover. Although the Air Force did not publicly release information on how many of the planes could be salvaged, the net worth of the planes at risk was $5.3 billion, more than three times the cost of building the base from scratch.[18] As John Conger, who oversaw installations for the Defense Department during the Obama administration, has warned, "There are plenty of other bases where the consequences of a direct hit are just huge."[19]

Beyond coastal military installations, other centers of military and political power also face potentially paralyzing threats from the impacts of climate change. Take flooding in the US capital, Washington, DC. In 2011, Bilal Ayyub, a professor of engineering at the University of Maryland, examined a scenario in which a strong hurricane—let's call it Hurricane Leo—hits the southeastern coast of Virginia and pushes massive amounts of water into the Chesapeake Bay.[20] His projections send a 16-foot (4.9 m) storm surge barreling up the Potomac River toward Washington, inundating large parts of the city and surrounding areas. If that hurricane also brought crippling rains, as Hurricane Harvey did in Houston, the resulting flood waters could shut down the headquarters of numerous federal agencies and reach the White House lawn. Severe flooding would also damage military installations around the city, including Fort McNair and the Navy and Joint Base Anacostia-Bolling, which is home to many military officers. Even if the buildings could still support operations, employees would have a hard time getting through the flooded streets and subway system to work.

Any number of seemingly minor failures in Washington's flood protection system could put parts of the city underwater.[21] Critical federal employees up and down the chain of command would have to brave flooded streets and transportation systems just to get to their jobs. Key services and decisions could be delayed, and the ability of the nation to respond to events in other parts of the country or the world could be impaired. Even local plans to evacuate might prove impossible to execute. Suppose our fictional hurricane had also hit Norfolk before it traveled the short 200 miles (322 km) northward on its way to Washington. Imagine the disruption that would follow the simultaneous crippling of both the US capital and the area that former US Secretary of Defense Leon Panetta once described as "perhaps the greatest concentration of military might in the world."[22] It's not far-fetched to envision hostile powers or groups exploiting the temporary chaos to wreak geopolitical mischief.

Another risk is that several simultaneous catastrophic events, or a series of disasters in rapid succession, could overtax the capacity of countries to maintain order. According to one study, if emissions continue unabated, by 2100 the world's population could face multiple climate hazards at the same time. In some places, that could mean as many as six climate-related disasters striking simultaneously.[23] As US intelligence services have recognized, multiple extreme weather events occurring within a small region or over a short period of time compound their impacts and undercut governments' abilities to cope. And when they occur in clusters or rapid succession, they can cause more damage than even a single powerful event.[24]

In the absence of increased resilience, some of the most powerful governments will find that their ability to project power and protect their citizens is threatened by climate change. Other great-power centers, including Beijing and London, also face significant risks from flooding. In China, researchers have developed models to identify areas with high concentrations of critical infrastructure that serves large populations, or infrastructural "hot spots." The models revealed that floods could jeopardize many of Beijing's critical services, including electricity, wastewater, and transportation.[25] Britain's Environment Agency has identified key buildings in London's government quarter that are at high risk of flooding; they include the Houses of Parliament, Whitehall, and City Hall. Much of London's critical infrastructure is also prone to flooding, including eight power stations, fifty-one railway stations, sixteen hospitals, and more than a thousand electricity substations.[26]

If climate-exacerbated disruptions are widespread and long lasting, they could create a power vacuum that allows geopolitical competitors to increase their spheres of influence or hostile nonstate actors to sow instability. Closer to home, climate change also undermines the traditional idea that the security of the great powers' capital cities and critical military hubs can be taken for granted. There is a third risk that must be considered, and it has to do with the impact of climate change on great-power competition.

THE NEW RACE FOR RESOURCES

Nations have always jockeyed for control of resources and economic opportunities, from raw materials and fertile land to access to markets. Climate change will exacerbate that competition, particularly with regard to food and water, and in the absence of cooperation agreements, it will create new sources of competition. The tensions over water between Pakistan and India, described earlier in this chapter, provide a glimpse of what may lie ahead. This climate-induced competition is especially obvious in the Arctic, where a geopolitical game has begun as the Arctic Ocean sea ice melts, opening navigation routes and unlocking access to rich natural resource deposits.

For most of recorded history, the world largely left the Arctic and its native peoples alone. The harsh, cold weather conditions and a frozen, inhospitable sea kept the global powers disinterested. But with climate change, the Arctic is warming quickly, opening the tantalizing prospect of its huge stores of untapped natural-resource deposits and shorter shipping routes. Beneath its landscape lie potentially vast tracts of minerals and precious metals, as well as an estimated 13 percent of the world's undiscovered oil and 30 percent of the world's undiscovered natural gas. Some of the world's largest fish stocks live here. Warmer temperatures have also opened new passageways for maritime exploration. Crossing the Arctic Ocean can cut the distance between Asian and European ports by some 30 percent or 40 percent compared to passage through the Suez Canal. The Arctic route saves both fuel and time. Whoever controls passage through the region would reap considerable economic benefits. Those with rights to the region's massive fossil-fuel deposits will also determine whether this carbon stays in the ground or further warms the earth.

Russia has already begun to make its play. With almost 15,000 miles of coastline running largely parallel to one of the two major Arctic shipping routes, Russia has sought to expand its presence in the region aggressively. In 2007, Russia sent two mini-submarines to the bottom of the Arctic Ocean, two and a half miles below the North Pole, planting a flag in a Hollywood-worthy ploy. Geopolitical competitors quickly ridiculed the gesture, but Russia has continued to invest relentlessly in the Arctic ever since. It has reactivated Soviet-era bases along the Northern Sea Route, conducted increasingly large military exercises in the region, and added to its fleet of more than forty icebreakers. In 2017, all this chest-puffing spurred the US-led North American Treaty Organization (NATO) to launch its largest war game since the end of the Cold War, pulling together 50,000 troops, 65 warships, and 250 planes to rehearse a possible response to "an act of armed aggression" in the Arctic.[27]

Russia's assertion of power in the Arctic faces significant competition, even from a country with no territory in the region. China began calling itself a "near-Arctic state" at least as early as 2012. That same year, the Chinese engaged in their own display of Arctic reach by sending a cargo ship, aptly named the Snow Dragon, to traverse the Arctic Ocean. Several years later, President Xi Jinping unveiled China's Arctic policy, making the country's intention to play a significant role in the region explicit. Within its Belt and Road Initiative, China has developed plans for a "polar silk road" linking China to Europe. According to estimates from China's Polar Research Institute, 5 to 15 percent of the country's trade could traverse the Arctic as early as 2020.[28]

Meanwhile, the United States is currently dead last in the footrace of nations competing for the Arctic. The US government simply has not made the necessary investments in infrastructure, communications, and ice-worthy ships to project dominant power in the Arctic region. In addition, because political wrangling has kept it from ratifying the UN Convention on the Law of the Sea, an international treaty governing the region, the United States has no seat at the table as the rest of the world's nations determine the international governance of the newly opening Arctic Ocean.

RESILIENCE PROTECTS NATIONAL SECURITY

When it comes to national security, how does a nation continue to project power while confronting the emerging security risks posed by the great disrupter that is climate change? How does a nation even identify those risks? Once it does, what can it do about them? It was not until the tail end of President Obama's second term that the US government tried to address these problems on a government-wide basis.

In the fall of 2015, the Obama White House assembled senior federal leaders to seek input on how best to tackle the growing national security risks posed by climate impacts. Federal agency officials initially greeted the effort with deep skepticism, questioning whether it was even worth their time. At the first meeting, held in a huge conference room on the fifth floor of the wedding-cake-like Old Executive Office Building next to the White House, many attendees sat with arms folded, leaning back deeply into their chairs. They expressed dismay at spending effort on a problem that seemed far-off in the future. "Why do we have to do this, if it's just a fad?" they asked.

The same reaction greeted Richard Holbrooke, President Obama's special envoy for Afghanistan and Pakistan, when he raised the issue of climate change at a meeting of national security leaders. When the conversation turned to the fractured relations between India and Pakistan, Holbrooke interjected, "There's a global warming dimension to this struggle, Mr. President."[29] Holbrooke feared that melting glaciers could cause massive flooding in Pakistan and possibly India, feeding a humanitarian crisis and exacerbating the security situation. The point was lost on some of his colleagues. After the meeting, several attendees reportedly wondered, "Was Holbrooke kidding?"[30]

But over time, as the nature and scope of the challenge came more clearly into focus, the mood inside the federal government changed. Discussions revealed that no single department or agency could manage the challenge alone; the challenge required streamlined coordination among scientists, the military, intelligence officers, and policymakers. When the National Security Council staff finally presented the twenty-some agencies with a proposed "Presidential Memorandum on Climate Change and National Security," each agency gave its unqualified approval and, not long afterward, the president signed it. The memorandum declared, for the first time ever, that it was the policy of the US government "to ensure that the current impacts of climate change, and those anticipated in the coming decades, be identified and considered in the development and implementation of relevant national security doctrine, policies, and plans."[31]

The memorandum went further by putting in place a framework for implementation, including establishing and bolstering requirements for detailed agency planning. Importantly, the document created a policy group made up of senior officials from multiple government agencies. The group would analyze the most urgent climate-related national security concerns, gather existing climate data scattered across the sprawling federal government, and make climate information available to intelligence experts and policymakers. The memorandum directed agencies to identify the countries and regions most vulnerable to climate change. Crucially, the document instructed the policy group to find ways the US government could assist other countries in their efforts to build resilience. It also gave the agencies a detailed work plan to consider areas of pressing concern, such as the implications of climate change for mass migration, global food and water security, global health security, economic strength, and critical infrastructure.

Shortly after he took office, President Trump revoked the historic memorandum, and the agencies never had a chance to develop that body of resilience planning. Some of the work continued in corners of his administration, but as climate impacts grow more severe, the US and other governments will need to devote more attention to the nexus between climate change and national security.

As we have seen, climate change is capable of destabilizing countries from within, creating spaces for violent, antidemocratic forces to challenge governments. It can undermine great-power centers and their military infrastructure and spawn new sources of power and competition among nations. To counter these upheavals, governments must first understand what is at stake, and that requires getting the best science and analysis into the hands of national security decision-makers. With that understanding, policymakers can then make strategic decisions about where and how to build resilience to blunt climate change's disruptive impact on national and global security.

PRESCRIPTIONS AND PROVOCATIONS

- The federal government should establish a system that provides national security policymakers with intelligence and risk assessments that are informed by the best-available climate science.
- Federal agencies should develop strategies to address climate-related risks in national security hotspots and other areas of strategic importance, including economic impacts, food and water insecurity, increased migrant flows, and public health threats.
- The Department of Defense and other relevant agencies should invest in measures to protect military assets to ensure readiness in the face of climate change, as well as screen all proposed military acquisitions for climate resilience. Defense contractors should identify and address climate-related risks to their supply chains and operations to avoid disruption in military procurement.

CONCLUSION

Silo-Breakers, Translators, and Communicators

The planes take off every day in perfectly synchronized succession from four heavily fortified bases. One is in the suburbs of Houston, one in a remote area of Uruguay, another outside of Cairo, and a fourth in Durban, South Africa. Despite the slick logo of the international consortium emblazoned on their sides, the aircraft look somewhat cartoonish. Specially designed for their unique mission, they have disproportionately large wings that connect to a narrow, stubby body just big enough for three people and cargo. The enormous wings help the craft stay aloft in the thin air at 65,000 feet. And four powerful engines, twice as many as a commercial airliner, propel the planes with their heavy cargo—a dense mass of molten sulfur.

The planes take off once or twice a day, every day of the year. Each craft flies five hundred missions annually; each mission takes about five hours to complete. More planes are added to the fleet every year. By the fifteenth year of the mission, there will be nearly a hundred planes making sixty thousand annual trips to the lower atmosphere, spraying a fine layer of sulfur particles that reflect solar radiation back into space. When the program was first launched, the cooling effects on the planet were almost instantaneous, despite the concentrated layer of greenhouse gases that surrounds the globe.

An international agreement governs the whole process, including how much sulfur is sprayed, when, and how. The trivial price tag, $3.5 billion in startup costs and $2.25 billion in annual operational costs, could be a rounding error in the government budgets of any of the bigger countries in the consortium. The cost-sharing arrangement among participating nations is purely for political symbolism. To be sure, the international agreement took time to negotiate, for it involved many more countries than those in which the bases are located, but fear drove the consensus. All life on the planet now depends on the program's orderly, continuous, and perpetual operation, and humanity's future hinges on the program working as intended.

This fictional anecdote is based on scholars' actual calculations of how this geoengineering effort might be undertaken.[1] It offers a realistic account of what a frequently proposed geoengineering program would look like, in this case, stratospheric aerosol injection, or SEI. Most advocates of SEI and similar geoengineering techniques point out that the world should resort to this technology only to complement an aggressive emissions-reduction program. They maintain that geoengineering is not a substitute for the necessary transition to a zero-emissions world. But it is very easy to see how the allure of the impossibly cheap short-term benefits of SEI would lull the world into complacency. Geoengineering solutions would provide countries with a powerful excuse to dodge the laborious, politically costly slog of cutting emissions and building resilience.

Other concerns make this an irresponsible choice carrying unacceptable risks. Because the complexities of the planet's climate system are not well understood, geoengineering's results and side effects may turn out very differently than what is expected. Indeed, the cure may prove as bad as the disease, if not worse. SEI could well result in more frequent tropical cyclones.[2] Studies of large volcano eruptions, which have effects comparable to SEI, suggest that the loss in solar radiation would mostly cancel out any crop-yield benefits induced by global cooling.[3]

In addition, SEI would not stop the acidification of the oceans. International conflict could arise if SEI wrought havoc with weather conditions in some regions of the world, even if it improved conditions elsewhere. And then there's the terrifying prospect referred to in the literature as "termination shock," the rapid and catastrophic spike in global temperatures that would swiftly follow should terrorism, war, natural disasters, political conflict, or some other eventuality interrupt the sulfur-spraying flights.

In short, SEI-style geoengineering cannot credibly substitute for cutting emissions and building resilience. Even if it were used as a complementary measure, the risks are enormous. Yet if we don't redouble our efforts on both emissions-cutting and resilience fronts, these are the kinds of desperate options with which we will be left. In this light, the need to put to work many of the lessons in this book becomes all the more pressing.

REBUILDING AFTER TRUMP

This book has not dwelled extensively on the backsliding on climate action in the United States that has occurred during the presidency of Donald Trump, but it is a reality we cannot leave out. Trump's announced withdrawal from the Paris Agreement captured media attention and remains his most salient "contribution" to the climate debate. But beyond the headlines, the Trump administration has inflicted substantial damage on the climate resilience agenda.

Some are acts of commission. The president axed the national flood risk-management standard, as well as the Presidential Memorandum on Climate Change and National Security. The Council on Climate Preparedness and Resilience, which brought together federal agencies to coordinate their response to climate impacts, and similar mechanisms simply stopped working. The current White House did its best to bury the findings of the Fourth National Climate Assessment, a massive, authoritative, and Congressionally mandated study of climate change impacts on the United States. Congress has largely blunted the draconian budget cuts to resilience-related research and national and international programs proposed by the Trump White House, but there has been little growth or innovation. Valuable time has gone to waste.

Others are acts of omission. This includes a vacuum of US climate leadership in international organizations and forums such as the G7 and G20, where efforts to promote climate resilience could have been more ambitious and influential. Meanwhile, the federal government has remained mostly silent on domestic climate resilience at a time when many states and cities hunger for guidance, support, and resources.

To be sure, Trump's rejection of the Paris Agreement jolted governors, mayors, CEOs, and others into action. The initiatives "We Are Still In" and "America's Pledge" emerged to encourage state and local governments, businesses, and civil society to do more so that the United States can meet its Paris Agreement commitments in the absence of federal action. But these efforts have focused almost exclusively on cutting greenhouse-gas emissions. Resilience must play a much larger role in these initiatives. At the same time, we must recognize that state and local action cannot fully substitute for the absence of an active federal government.

During the Trump years, more Americans than ever before started telling pollsters that they feel personally affected by climate change. Survey results from December 2018 showed that about seven in ten Americans are least "somewhat worried" about global warming, and three in ten say they are "very worried," a record high since this polling started in 2008.[4] About half (48 percent) of the American public told pollsters that global warming is harming people in the United States "right now"—a sixteen-point increase since March 2015.[5] Shaken by extreme weather events, the American public appears more receptive than ever to resilience messages. This should provide an opening for political and business leaders who can offer clear, practical, and action-oriented approaches to support communities trying to cope with the impacts of climate change. The Green New Deal, an ambitous vision for climate action proposed in the wake of the 2018 US midterm elections, prominently included resilience as a key goal.

When the current political moment shifts, many of the innovations launched during the Obama administration and described in this book should be brought back, updated, and expanded. The to-do list of resilience opportunities left behind by outgoing Obama officials offers a good place to start.[6] In addition, Americans will find that while the US federal government slumbered during President Trump's tenure, the rest of the world kept moving forward on resilience, creating rich reservoirs of international learning and experience the United States should tap.

SILO-BREAKERS, TRANSLATORS, AND COMMUNICATORS

As we were completing this book, we were struck by several themes that came through clearly across the ten lessons. The first is that valuable insights about resilience are often trapped inside specialized communities, each with its own technical language, its own professional biases, and its own limited audience. This hinders learning and progress. Promoting resilience will require breaking out of these silos, deploying many different tools across disciplines and sectors, and developing a common language for resilience. Silo-breakers play a critical role.

As this project took shape, we encountered many silo-breakers. Silo-breakers are people who can reach across professional and academic disciplines, connect different groups of people, and drive collaborative solutions. They include Ann Phillips, whom we met in the introduction and who is bringing together stakeholders from different professional worlds in search of answers for Norfolk's flooding challenge. They also include Dean Linda Fried of Columbia University's Mailman School of Public Health, who is trying to bring climate change into medical and public health curricula. Other silo-breakers are Daniel Kreeger, who is seeking to persuade architecture schools to teach their students about climate change, and Mark Carney, the UK central bank governor who has pushed his colleagues to think outside their box.

A second theme is that we must invest heavily in tools and knowledge to make better decisions about the future in a warming world. Whether it's corporations protecting their supply chains, mental health professionals treating patients, farmers deciding what to plant and when to plant it, or engineers siting and designing infrastructure, all stakeholders need a much better grasp of what a future under climate change will mean for their occupations and livelihoods. To that end, climate change impacts cannot remain abstractions, or something too complex to be understood. That's why we need translators. Translators are people who can turn complex climate data and information into language that speaks directly to the practical problems people are trying to solve in their daily lives in the face of climate disruptions.

We have written about several translators in this book. They include Planet's Will Marshall, who is seeking to bring minisat data to millions of users in a format they can easily understand. They include Ahmad Wani of One Concern, who is working to create hyperlocalized risk models for businesses and governments, and Julian Ramirez-Villegas of the International Center for Tropical Agriculture, who explains to Colombian farmers how to use climate data to make planting decisions.

In the course of writing this book, we also gained a deeper appreciation for great communicators. These are people who can persuasively convey the opportunities and benefits that resilience brings, not only tomorrow but also today. That's the message that will be essential for overcoming inertia and selling resilience action to the public. Communicators are also people who can speak with officials and the public about climate issues that may often seem overwhelming and help them to see a path forward.

Among the great communicators we met or heard from were former Mayor of Hoboken Dawn Zimmer, who persuaded her city to invest in resilience, and Henk Ovink, who speaks eloquently about how to apply Dutch building concepts resiliently to modern problems. They also include Christine Morris and Susanne Torriente, chief resilience officers of Norfolk and Miami-Dade County, respectively. Both are passionate advocates of resilience and understand how to get officialdom and the larger community to address this urgent challenge.

To make progress, we will need many, many more silo-breakers, translators, and communicators. We must redouble our efforts to recruit and empower the next generation.

LOOKING AFTER THE MOST VULNERABLE FIRST

Finally, a theme that emerged again and again is that building resilience means protecting the most vulnerable communities and members of society first. This will demand identifying who they are, where they live, and how climate impacts them today and how it may impact them tomorrow. It calls for bringing their interests to the table when making decisions about investing in resilience, and it means protecting the vital services on which they rely, such as public transportation and health care facilities. It also demands thinking about who will be left in the lurch if insurance companies drop coverage, or bond markets shut them out, or a storm leaves them homeless. Finally, protecting the most vulnerable from climate impacts requires building systems to provide them with information, money, and other support. Whether societies are successful in coping with the impacts of climate change will largely depend on whether they are able to protect those least able to weather the upheaval.

As we were finishing the book, we revisited Norfolk to see what had happened since we were in government and had received the officials' worried calls. What we found gave us cause for hope. Norfolk and the surrounding cities had come together and agreed on common sea-level rise scenarios so that all the stakeholders can plan and build against the same benchmark. Even the city of Virginia Beach, which had historically opposed higher-end projections, had reversed its position and become a staunch advocate of aggressive action, partly in reaction to Hurricane Matthew in 2016. Meanwhile, Virginia's governor established a cabinet-level position to advise him on coastal protection, and in the Norfolk area, a committee of elected officials convened to develop responses to flooding in the region.

The climb is steep, and many obstacles, large and small, remain. Federal funds are hard to come by, and Norfolk still struggles with how to cover the huge costs of resilience. Poor communities need to be relocated, but where they will resettle is uncertain, as are the opportunities that will await them. Elderly residents often decline to participate in buyout programs. Some are emotionally attached to their homes and don't want to move; others don't want to live in elevated structures with steep flights of stairs.

Yet Norfolk's leaders, businesses, and citizens are fully aware of the challenge. They understand what needs to be done and are immersed in devising solutions. Now it's up to all of us to help them and the thousands of other communities in the United States and around the world that face this shared existential challenge.

NOTES

Introduction

1. David Kramer, "Norfolk: A Case Study in Sea-Level Rise," *Physics Today* 69, no. 5 (2016): 22–25.
2. Kramer, "Norfolk."
3. UN Environment Programme, "Emissions Gap Report 2018" (Nairobi, Kenya: United Nations Environment Programme, 2018).

Chapter 1

1. Ed Rappaport, "Preliminary Report: Hurricane Andrew 16–28 August, 1992" (Miami, FL: National Hurricane Center, 1993).
2. Rappaport, "Preliminary Report."
3. Peter Applebome, "After the Storm; Amid the Fallen Buildings, a Host of Questions about How They Were Built," *New York Times*, September 6, 1992.
4. Dade County Grand Jury, "Final Report of the Dade County Grand Jury" (Miami: Eleventh Judicial Circuit Court of Florida, 1992) 2.
5. Juliet Christian-Smith, "A Bridge over Troubled Waters: How the Bay Bridge Was Rebuilt without Considering Climate Change," *Union of Concerned Scientists Blog*, February 6, 2015; Esmé E. Deprez and Martin Z. Braun, "NYC Subway-Station-Turned-Fish-Tank Poses $600 Million Dilemma," *Bloomberg*, December 5, 2012.

6. John McQuaid, "Dutch System of Flood Control an Engineering Marvel," Nola.com, November 13, 2005.

7. Michael Kimmelman, "The Dutch Have Solutions to Rising Seas. The World Is Watching," *New York Times*, June 15, 2017.

8. Multihazard Mitigation Council, *Natural Hazard Mitigation Saves: An Independent Study to Assess the Future Savings from Mitigation Activities*, vol. 1: *Findings, Conclusions, and Recommendations* (Washington, DC: Multihazard Mitigation Council, 2005).

9. Multihazard Mitigation Council, "Natural Hazard Mitigation Saves: 2017 Interim Report" (Washington, DC: Multihazard Mitigation Council, 2017).

10. C. M. Shreve and I. Kelman, "Does Mitigation Save? Reviewing Cost-Benefit Analyses of Disaster Risk Reduction," *International Journal of Disaster Risk Reduction* 10, part A (2014), 213–235.

11. Multihazard Mitigation Council, "Natural Hazard Mitigation Saves: 2018 Interim Report" (Washington, DC: Multihazard Mitigation Council, 2018).

12. Insurance Institute for Business and Home Safety, "Rating the States: 2018 Atlantic and Gulf Coast States an Assessment of Residential Building Code and Enforcement Systems for Life Safety and Property Protection in Hurricane-Prone Regions" (Tampa, FL: Insurance Institute for Business and Home Safety, 2018).

13. Federal Emergency Management Agency, "Federal Flood Risk Management Standard" (Washington, DC: Federal Emergency Management Agency, 2015).

14. Houston Public Works, "Floodplain Management Data Analysis: Chapter 19" (Houston, TX: City of Houston, 2018).

15. Stephen L. Quarles and Kelly Pohl, "Building a Wildfire-Resistant Home: Codes and Costs" (Bozeman, MT: Headwater Economics, 2018).

16. Nick Perry, "Air Force Ignored Rising-Sea Warnings at Radar Site," *Air Force Times*, October 18, 2016.

17. Perry, "Air Force Ignored."

18. Curt D. Storlazzi, Stephen Gingerich, Peter Swarzenski, Olivia Cheriton, Clifford Voss, Ferdinand Oberle, Joshua Logan, Kurt Rosenberger, Theresa Fregoso, Sara Rosa, Adam Johnson, Li Erikson, Don Field, Greg Piniak, Amit Malhotra, Mark Finkbeiner, Ap van Dongeran, Ellen Quataert, Arnold van Rooijen, Edwin Elias, Mattijs Gawehn, Annamalai Hariharasubramanian, Matthew Widlansky, Jan Hafner, and Chunxi Zhang,, "The Impact of Sea-Level Rise and Climate Change on Department of Defense Installations on Atolls in the Pacific Ocean (RC-2334)" (Reston, VA: US Geological Survey, 2017).

19. R. S. Kafalenos, K. J. Leonard, D. M. Beagan, V. R. Burkett, B. D. Keim, A. Meyers, D. T. Hunt, R. C. Hyman, M. K. Maynard, B. Fritsche, R. H. Henk, E. J. Seymour, L. E. Olson, J. R. Potter, and M. J. Savonis, "What Are the Implications of Climate Change and Variability for Gulf Coast Transportation?," in *Impacts of Climate Change and Variability on Transportation*

Systems and Infrastructure: Gulf Study, Phase I. A Report by the US Climate Change Science Program and the Subcommittee on Global Change Research. Final Report of Synthesis and Assessment Product 4.7, ed. M. J. Savonis, V. R. Burkett, and J. R. Potter (Washington, DC: US Department of Transportation, 2008).

20. Christina Nunez, "As Sea Levels Rise, Are Coastal Nuclear Plants Ready?," *National Geographic*, December 16, 2015.

21. Union of Concerned Scientists, "Flood Risk at Nuclear Power Plants," Union of Concerned Scientists (website), n.d. https://www.ucsusa.org/nuclear-power/nuclear-power-accidents/flood-risk-at-nuclear-power-plants.

22. Daniel Kreeger, "Remarks at Institutionalizing Climate Change into Decision-Making" (speech, Baruch College, New York, NY, September 26, 2018).

23. Bilal Ayyub, ed., *Climate Resilient Infrastructure: Adaptive Design and Risk Management* (Reston, VA: American Society of Civil Engineers, 2018); Richard Dial, Bruce Smith, and Gheorghe Rosca Jr., "Evaluating Sustainability and Resilience in Infrastructure: Envision, SANDAG and the LOSSAN Rail Corridor" (paper presented at the 2014 International Conference on Sustainable Infrastructure, Long Beach, CA, November 6–8, 2014).

24. Samuel Fankhauser and Raluca Soare, "A Strategic Approach to Adaptation in Europe" (London: Centre for Climate Change Economics and Policy and Grantham Research Institute on Climate Change and the Environment, 2012).

25. Michael Slezak, "Australian Inferno Previews Fire-Prone Future," *New Scientist*, January 16, 2013.

26. Lauren Sommer, "As California's Population Grows, People Are Moving into More Fire-Prone Areas," *All Things Considered*, National Public Radio, October 27, 2017.

27. Bettina Boxall, "A Simple but Seldom-Used Tactic to Prevent Wildfires: Turn Off the Power Grid When Winds Pick Up," *Los Angeles Times*, November 24, 2017.

28. Don Thompson, "Official: California Must Mull Home Ban in Fire-Prone Areas," *Sacramento Bee*, December 11, 2018.

29. Kerry Sanders, "'Dome Home' Weathers Storm," *NBC News*, September 16, 2004.

30. Michael South, "How Much Does a Monolithic Dome Home Cost?," *Monolithic Dome Institute Blog*, March 20, 2013.

31. Jonathan Alarcon, "The FLOAT House—Make it Right / Morphosis Architects," *Arch Daily*, August 2, 2012.

32. The Contemporist, "Flood-Proof House by Studio Peek Ancona," *Contemporist*, August 27, 2010.

33. Ashleigh Davis, "Blooming Bamboo Home by H&P Architects," *Deezeen*, September 25, 2013.

34. Siddharth Narayan, Michael W. Beck, P. Wilson, C. Thomas, A. Guerrero, C. Shephard, B. G. Reguero, G. Franco, C. J. Ingram, and D. Trespalacios, "Coastal

Wetlands and Flood Damage Reduction: Using Risk Industry-Based Models to Assess Natural Defenses in the Northeastern USA" (London: Lloyd's Tercentenary Research Foundation, 2016).

35. Environmental Protection Agency, "Reducing Stormwater Costs through Low Impact Development (LID) Strategies and Practices" (Washington, DC: US Environmental Protection Agency, 2007).

36. Helen Roxburgh, "China's 'Sponge Cities' Are Turning Streets Green to Combat Flooding," *The Guardian*, December 27, 2017.

37. The Partnership for Water Sustainability in British Columbia, "Sponge Cities," *Waterbucket*, May 5, 2018.

38. Faith Ka Shun Chan, James A. Griffiths, David Higgitte, Shuyang Xua, Fangfang Zhug, Yu-Ting Tang, Yuyao Xu, and Colin R. Thorne, "'Sponge City' in China—a Breakthrough of Planning and Flood Risk Management in the Urban Context," *Land Use Policy* 76 (2018): 772–778.

Chapter 2

1. Pepe's restaurant was destroyed in a fire in August 2013.

2. Native Village of Kivalina v. ExxonMobil Corp., 696 F.3d 849 (9th Cir. 2012).

3. Kirsten Engel and Jonathan Overpeck, "Adaptation and the Courtroom: Judging Climate Science," *Michigan Journal of Environmental & Administrative Law* 3, no. 1 (2013): 4.

4. Order and opinion denying defendants' motion to dismiss at 3, Juliana et al. v. United States, No. 6:15-cv-01517-TC (D. Or. Nov. 10, 2016).

5. Story Hinckley, "Pakistani Farmer Sues Government to Curb Climate Change," *Christian Science Monitor*, November 16, 2015.

6. Robert McCoppin, Lisa Black, and Dan Hinkel, "Insurers Sue Chicago-Area Towns in Bid to Get Flood Money," *Chicago Tribune*, May 14, 2014.

7. Complaint at 20, Illinois Farmers Insurance Co. v. Metropolitan Water Reclamation District of Greater Chicago, No. 2014CH06608 (Ill. Cir. Ct. Apr. 16, 2014).

8. Robert McCoppin, "Insurance Company Drops Suits over Chicago-Area Flooding," *Chicago Tribune*, June 3, 2014.

9. John Echeverria, "Texas, Flooding, and Public Use," *Takings Litigation*, July 5, 2016.

10. Harris County Flood Control District v. Kerr, No. 13-0303, 2015 Tex. LEXIS 545 (Super. Ct. Jun. 12, 2015).

11. Burgess v. Ontario Minister of Natural Resources and Forestry, No. 16-1325 CP (Super. Ct. J. filed Sept. 14, 2016).

12. Jennifer Klein, "Potential Liability of Governments for Failure to Prepare for Climate Change" (New York: Sabin Center for Climate Change Law, 2015), 26.

13. Meghna Chakrabarti, "Mayor Bill De Blasio on Why New York City Is Suing 5 Major Oil Companies," *Here & Now*, January 11, 2018.

14. Edward-Isaac Dovere, "Schwarzenegger to Sue Big Oil for 'First Degree Murder,'" *Politico*, March 12, 2018.

15. John Schwartz, "Judge Throws out New York Climate Lawsuit," *New York Times*, July 19, 2018.

16. Anne C. Mulkern, "Here's the Strategy behind Cities' Lawsuits against Big Oil," *E&E News*, March 28, 2018.

17. Agence France-Presse, "German Court to Hear Peruvian Farmer's Climate Case against RWE," *The Guardian*, November 30, 2017.

18. Complaint and jury demand, Conservation Law Foundation, Inc. v. Shell Oil Products US et al., No. 1:17-cv-00396 (D.R.I. Aug. 8, 2017).

19. Jon Chesto, "Conservation Law Foundation Sues Exxon Mobil," *Boston Globe*, September 29, 2016.

20. Sophie Marjanac, Lindene Patton, and James Thornton, "Acts of God, Human Influence and Litigation," *Nature Geoscience* 10 (2017): 616.

21. Noel Hutley and Sebastan Hatford Davis, "Climate Change and Directors' Duties: Memorandum of Opinion" (Sydney, Australia: Center for Policy Development and the Future Business Council, 2016).

22. S. C. Herring et al., eds., "2018: Explaining Extreme Events of 2016 from a Climate Perspective," *Bulletin of the American Meteorological Society* 99, no. 1 (2018): S1-S157.

23. Jessica Wentz, "Government Officials' Liability after Extreme Weather Events: Recent Developments in Domestic and International Case Law," *Climate Law Blog*, April 9, 2015.

24. Agence-France Press, "French Mayor Jailed over Floods That Left 29 Dead," *The Local France*, December 12, 2014.

25. Peter Guinta, "County: Old A1A Lawsuit Settled," *St. Augustine Record*, January 18, 2013.

26. Jordan v. St. Johns County, No. 5D09-2183, 5D09-4378 & 5D09-4379, 2011 Fla. App. LEXIS 7253 (Fla. Dist. Ct. App. 5th Dist., May 20, 2011).

27. J. Wylie Donald, "Harvey Cedars v. Karan: Condemnation at the Shore and the Evolution of the Common Law," *Lexology*, July 29, 2013.

28. *Borough of Harvey Cedars v. Karan*, 214 N.J. 384 (2013), at 389.

29. Mary Ann Spoto, "Harvey Cedars Couple Receives $1 Settlement for Dune Blocking Ocean View," *NJ.com*, September 25, 2013.

30. Turek v. Zoning Bd. of Appeals, No. LNDCV156063404S, 2018 Conn. Super. LEXIS 778 (Super. Ct. Apr. 4, 2018), at 15.

31. Meagan Flynn, "Federal Judge Rules Extreme Texas Prison Heat Is Cruel and Unusual Punishment," *Houston Press*, July 20, 2017.

32. Cole v. Collier, No. 4:14-CV-1698, 2017 U.S. Dist. LEXIS 112095 (S.D. Tex. July 19, 2017), at 102.

33. UN Environment Programme, "The Status of Climate Change Litigation: A Global Review" (Nairobi, Kenya: UN Environment Programme, 2017).

34. City of Oakland v. BP P.L.C., No. C 17-06011 WHA; No. C 17-06012 WH, 2018 U.S. Dist. LEXIS 106895 (N.D. Cal. June 25, 2018), at 31.

35. Ibid., 24.

36. Schwartz, "Judge Throws out New York Climate Lawsuit."

37. Juliana v. United States, 217 F. Supp. 3d 1224 (D. Or. 2016).

38. Ibid., 52.

Chapter 3

1. Mark Carney, "Breaking the Tragedy of the Horizon: Climate Change and Financial Stability" (speech, Bank of England, London, September 29, 2015).

2. Ibid.

3. Arthur Levitt, "Investor Education: Disclosure for the 1990s" (speech, University of Virginia, Charlottesville, November 1, 1995).

4. US Securities and Exchange Commission, "Fast Answers: The Laws That Govern the Securities Industry," US Securities and Exchange Commission, last modified, October 1, 2013, https://www.sec.gov/answers/about-lawsshtml.html.

5. Masahiko Haraguchi and Upmanu Lall, "Flood Risks and Impacts: A Case Study of Thailand's Floods in 2011 and Research Questions for Supply Chain Decision Making," *International Journal of Disaster Risk Reduction* 14, part 3 (2015): 256–272.

6. Steve Messner, L. Moran, Gregory Reub, and J. Campbell, "Climate Change and Sea Level Rise Impacts at Ports and a Consistent Methodology to Evaluate Vulnerability and Risk," *WIT Transactions on Ecology and the Environment* 169 (2013): 141–153.

7. Aaron Kressig, Logan Byers, Johannes Friedrich, Tianyi Luo, and Colin McCormick, "Water Stress Threatens Nearly Half the World's Thermal Power Plant Capacity," *Insights: World Resources Institute Blog*, April 11, 2018.

8. Task Force on Climate-Related Financial Disclosures, "Final Report: Recommendations of the Task Force on Climate-Related Financial Disclosures" (Basel, Switzerland: Task Force on Climate-Related Financial Disclosures, June 2017).

9. Stanislas Dupré, "Exploring Tragedy of the Horizon" (remarks at Two-Degrees Investing Initiative event, New York City, September 21, 2016).

10. European Commission, "Sustainable Finance: Commission's Action Plan for a Greener and Cleaner Economy," news release, March 8, 2018.

11. Lori Montgomery, "In Norfolk, Evidence of Climate Change Is in the Streets at High Tide," *Washington Post*, May 31, 2014.

12. Bob Buhr, Ulrich Volz, Charles Donovan, Gerhard Kling, Victor Murinde, and Natalie Pullin, "Climate Change and the Cost of Capital in Developing Countries: Assessing the Impact of Climate Risks on Sovereign Borrowing Costs" (London: Imperial College Business School and SOAS University of London, 2018) iv.

13. Christopher Flavelle, "Rising Seas May Wipe Out Jersey Towns and Millions in AAA Bonds," *Bloomberg*, May 25, 2017.

14. JLL and LaSalle, "Global Real Estate Transparency Index 2018" (Chicago: JLL and LaSalle, 2018).

15. Zillow Group, "Zillow Group Investor Relations," Zillow Group, http://investors.zillowgroup.com/.

16. Francesc Ortega and Suleyman Taşpinar, "Rising Sea Levels and Sinking Property Values: Hurricane Sandy and New York's Housing Market," *Journal of Urban Economics* 16 (2018): 81–100.

17. Asaf Bernstein, Matthew Gustafson, and Ryan Lewis, "Disaster on the Horizon: The Price Effect of Sea Level Rise," *Journal of Financial Economics* (Forthcoming).

18. Maarten Bosker, Harry Garretsen, Gerard Marlet, and Clemens van Woerkens, "Nether Lands: Evidence on the Price and Perception of Rare Natural Disasters," *Journal of the European Economic Association* jvy002 (2018): 413–453.

19. Jesse M. Keenan, Thomas Hill, and Anurag Gumber, "Climate Gentrification: From Theory to Empiricism in Miami-Dade County, Florida," *Environmental Research Letters* 13 (2018): 054001.

20. Bernstein, Gustafson, and Lewis, "Disaster on the Horizon."

21. Miami-Dade County, "Property Sale Disclosure," Miami-Dade County, https://www.miamidade.gov/environment/flood-disclosure.asp.

22. Stefanos Chen, "Buildings Rise in Flood Zones," *New York Times*, July 6, 2018.

23. Ray Lehmann, "Congress Let NFIP Off Hook for $16B Debt, Despite Less Than $10B in Claims," *Insurance Journal*, July 9, 2018.

24. Jenny Anderson, "Outrage as Homeowners Prepare for Substantially Higher Flood Insurance Rates," *New York Times*, July 28, 2013.

25. Insurance Journal, "Rep. Waters, Author of Flood Reform Act, Calls for Delay in Implementation," *Insurance Journal*, September 30, 2013.

26. David Jones, "Remarks at Institutionalizing Climate Change into Decision-Making" (speech, Baruch College, New York, NY, September 26, 2018).

27. Mary Williams Walsh, "How Wildfires Are Making Some California Homes Uninsurable," *New York Times*, November 20, 2018.

28. Insurance Institute for Business and Home Safety, "Structures at Alabama's Gulf State Park Are First-Ever IBHS FORTIFIED Commercial™–Hurricane Bronze Resilient Buildings," news release, November 2, 2018.

29. Jones, "Remarks at Institutionalizing Climate Change into Decision-Making."

Chapter 4

1. Names Redacted, "Emergency Supplemental Appropriations Legislation for Disaster Assistance: Summary Data," CRS Report No. RL33226 (Washington, DC: Congressional Research Service, 2008).

2. Gene L. Dodaro, "Progress on Many High-Risk Areas, While Substantial Efforts Needed on Others," report GAO-17-375T (Washington, DC: Government Accountability Office, 2017).

3. Leonardo Martinez-Diaz, "Investing in Resilience Today to Prepare for Tomorrow's Climate Change," *Bulletin of the Atomic Scientists* 74, no. 2 (2018): 66–72.

4. Martinez-Diaz, "Investing in Reselience."

5. Kathleen McGrory, "Despite Criticism, Miami Mayor Tomás Regalado Maintains Support," *Miami Herald*, January 20, 2013.

6. David Smiley, "How 4 A.M. Chats Persuaded Miami's Republican Mayor to Care about Sea-Level Rise," *Miami Herald*, October 6, 2017.

7. Republic of Fiji, "How Do We Get There? Submission to the Talanoa Dialogue" (Bonn, Germany: Secretariat of the United Nations Framework Convention on Climate Change, 2018).

8. Giancarlo Falcocchio, Awais Malik, and Constantine E. Kontokosta, "A Data-Driven Methodology for Equitable Value-Capture Financing of Public Transit Operations and Maintenance," *Transport Policy* 66 (2018): 107–115.

9. Vincent Viguié and Stéphane Hallegatte, "Urban Infrastructure and Rent-Capture Potentials," Policy Research Working Paper 7067 (Washington, DC: World Bank, 2014).

10. Eike Karola Velten, Matthias Duwe, Elizabeth Zelljadt, Nick Evans, and Marius Hasenheit, "Smart Cash for the Climate: Maximising Auctioning Revenues from the EU Emissions Trading System" (Berlin: Ecologic Institute and World Wildlife Fund, 2016).

11. California Climate Investments, "Annual Report to the Legislature on California Climate Investments Using Cap-and-Trade Auction Proceeds" (Sacramento: California Climate Investments, 2017).

12. Regional Greenhouse Gas Initiative, "The Investment of RGGI Proceeds through 2014" (New York: Regional Greenhouse Gas Initiative, 2016).

13. Robinson Meyer, "Will Washington State Voters Make History on Climate Change?," *The Atlantic*, August 18, 2018.

Chapter 5

1. S. M. Greenfield and W. W. Kellogg, "Inquiry into the Feasibility of Weather Reconnaissance From a Satellite Vehicle" (Santa Monica, CA: RAND Corporation, 1951).
2. Craig Fugate and Alice Hill, "The Small Towns Trump Is Hurting by His Denial of Climate Change," *Newsweek*, April 5, 2017.
3. Ashlee Vance, "The Tiny Satellites Ushering in the New Space Revolution," *Bloomberg*, June 29, 2017.
4. Christopher E. Schubert, Ronald Busciolano, Paul P. Hearn, Jr., Ami N. Rahav, Riley Behrens, Jason Finkelstein, Jack Monti, Jr., and Amy E. Simonson, "Analysis of Storm-Tide Impacts from Hurricane Sandy in New York," Scientific Investigations Report 2015–5036 (Reston, VA: US Geological Survey, 2015).
5. National Environmental Satellite, Data, and Information Service, "Data Dive: Five NOAA Databases that are Worth Exploring," *NESDIS News & Articles*, August 15, 2018 (the authors' calculations).
6. Shawna Wolverton, "Making the Move from Imagery to Insights with Planet Analytics," *Planet News*, July 18, 2018 (the authors' calculations).
7. World Meteorological Organization, "How Weather Services Reduce Disaster Risk: From Early Warnings to Climate Resilience" (Geneva: World Meteorological Organization, 2015).
8. JLT Re, "Catastrophe Models: In the Eye of the Storm" (London: JLT Re, 2018).
9. *The Economist*, "Software Can Model How a Wildfire Will Spread," *The Economist*, August 2, 2018.
10. Full disclosure: At the time of writing, Alice Hill served on One Concern's advisory board.
11. Roger Handberg, *The Future of the Space Industry: Private Enterprise and Public Policy* (Westport, CT: Greenwood, 1995), 60; US Congress, House of Representatives, Land Remote Sensing Commercialization Act of 1984, HR 5155, 98th Cong., 2nd sess., introduced in House March 15, 1984.
12. C. P. Williams and R. P. Mroczynski, "Impact of Commercial Remote Sensing: The American Experience," *Technology in Society* 11 (1989): 15–28; Tony Reichhardt, "Research to Benefit from Cheaper Landsat Images," *Nature* 400 (1999): 702.
13. Peter Folger, "Landsat: Overview and Issues for Congress," CRS Report No. R40594 (Washington, DC: Congressional Research Service, 2014), 5; US Congress, House of Representatives, Land Remote Sensing Policy Act of 1992, HR 6133, 102nd Cong., 2nd sess., introduced in House October 5, 1992.
14. James Lay, "Catastrophe Modelling: Out in the Open," *The Actuary*, September 6, 2018.

15. Risk Modelling Steering Group, "CatRiskTools: A Global Catalogue of Catastrophe Risk Analysis Tools" (Insurance Development Forum, 2018).

16. Catherine Vaughan, Suraje Dessai, Chris Hewitt, Walter Baethgen, Rafael Terra, and Mercedes Berterretche, "Creating an Enabling Environment for Investment in Climate Services: The Case of Uruguay's National Agricultural Information System," *Climate Services* 8 (2017): 62–71.

17. Shoumojit Banerjee, "IMD's New Weather Model Will Make Forecasts More Reliable," *The Hindu*, February 13, 2017.

18. US Government Accountability Office, "Climate Information: A National System Could Help Federal, State, Local, and Private Sector Decision Makers Use Climate Information," report GAO-16-37 (Washington, DC: U.S. Government Accountability Office, 2015).

19. Ibid., 15.

20. Ibid.

21. Michael Lewis, *The Undoing Project: A Friendship That Changed the World* (New York: Allen Lane, 2016).

22. Full disclosure: At the time of writing, Leonardo Martinez-Diaz worked for WRI.

23. Paul Reig, Tien Shiao, Roy Owens, and David Palochko, "Case Study: Aqueduct Informs Owens Corning Corporate Water Strategy" (Washington, DC: World Resources Institute, 2013).

24. Ibid., 1.

25. Maria Eliza Villarino, "Climate Services for Smarter Farming—What's It All About?," *CIAT Blog*, October 5, 2017.

26. Ibid.

27. Council on Climate Preparedness and Resilience, "Opportunities to Enhance the Nation's Resilience to Climate Change" (Washington, DC: White House, 2016).

Chapter 6

1. City of Cape Town Media Office, "City Continues Proactive Water Demand Management and Medium-to-Long Term Planning," news release, February 27, 2017.

2. Jenna Etheridge, "De Lille Warns 'Day Zero' for Cape Town's Municipal Water Supply Is March 2018," *News24*, October 4, 2017.

3. David B. LaFrance, "Day Zero, Defeat Day Zero," *Journal – American Water Works Association* 110, no. 3 (2018): 10.

4. Grant Smith and Martine Visser, "Behavioural Nudges as a Water Savings Strategy: Report to the Water Research Commission," WRC Report No. 2091/1/13 (Pretoria, South Africa: Water Research Commission, 2014).

5. Robyn Dixon, "How Cape Town Found Water Savings California Never Dreamed Of," *Los Angeles Times*, April 1, 2018.

6. David Halpern, *Inside the Nudge Unit: How Small Changes Can Make a Big Difference* (London: Ebury, 2015).

7. Exec. Order No. 13707, "Using Behavioral Science Insights to Better Serve the American People," 3 C.F.R. 56365 (2015).

8. Zeina Afif, "'Nudge Units'—Where They Came from and What They Can Do," *World Bank Let's Talk Development Blog*, October 25, 2017.

9. Anthony Leiserowitz, Edward Maibach, Connie Roser-Renouf, Seth Rosenthal, Matthew Cutler, and John Kotcher, "Climate Change in the American Mind: March 2018" (New Haven, CT: Yale Program on Climate Change Communication, 2018).

10. National Commission on Terrorist Attacks upon the United States, "The 9/11 Commission Report: Final Report of the National Commission on Terrorist Attacks upon the United States" (Washington, DC: National Commission on Terrorist Attacks upon the United States, 2004), 346.

11. Ibid., 344.

12. Richenda Connell, John Firth, Alastair Baglee, Anna Haworth, Jennifer Steeves, Caroline Fouvet, and Robin Hamaker-Taylor, "Navigating a New Climate – Assessing Credit Risk and Opportunity in a Changing Climate: Outputs of a Working Group of 16 Banks Piloting the TCFD Recommendations Part 2: Physical Risks and Opportunities" (Geneva: UN Environment Programme Finance Initiative, 2018).

13. Shaun Donovan, "Remarks at Too Big: Rebuild by Design: Transformative Response to Climate Change" (speech, National Building Museum, Washington, DC, November 19, 2018).

14. National Weather Service, "Service Assessment: Hurricane Katrina August 23-31, 2005" (Silver Spring: National Oceanic and Atmospheric Administration, 2005), 18.

15. Barry S. Goldsmith, D. W. Sharp, P. Santos, R. J. Ricks, Jr., and M. J. Moreland, "From the "Statement Heard Around the World" to Hurricane Threats and Impacts: The Evolution of Communicating Potential Impacts and Safety Messages since Katrina" (paper presented at the Special Symposium on Hurricane Katrina: Progress in Leveraging Science, Enhancing Response and Improving Resilience, New Orleans, LA, January 12, 2016).

16. Rebecca E. Morss et al., "Understanding Public Hurricane Evacuation Decisions and Responses to Forecast and Warning Messages," *Weather and Forecasting* 31 (2016): 395–417.

17. Dawn Zimmer, "Remarks at Too Big: Rebuild by Design: Transformative Response to Climate Change" (speech, National Building Museum, Washington, DC, November 19, 2018).

18. Paul J. Ferraro and Michael K. Price, "Using Non-Pecuniary Strategies to Influence Behavior: Evidence from a Large Scale Field Experiment," NBER Working Paper No. 17189 (Cambridge: National Bureau of Economic Research, 2011).

19. Kerri Brick, Samantha De Martino, and Martine Visser, "Behavioural Nudges for Water Conservation: Experimental Evidence from Cape Town" (Gothenburg, Sweden: Environment for Development Initiative, 2017).

20. Afif, " 'Nudge units.' "

Chapter 7

1. Donald G. McNeil Jr., "Zika Virus, a Mosquito-Borne Infection, May Threaten Brazil's Newborns," New York Times, December 28, 2015.

2. Megan R. Reynolds, Abbey M. Jones, Emily E. Petersen, Ellen H. Lee, Marion E. Rice, Andrea Bingham, Sascha R. Ellington, Nicole Evert, Sarah Reagan-Steiner, Titilope Oduyebo, Catherine M. Brown, Stacey Martin, Nina Ahmad, Julu Bhatnagar, Jennifer Macdonald, Carolyn Gould, Anne D. Fine, Kara D. Polen, Heather Lake-Burger, Christina L. Hillard, Noemi Hall, Mahsa M. Yazdy, Karnesha Slaughter, Jamie N. Sommer, Alys Adamski, Meghan Raycraft, Shannon Fleck-Derderian, Jyoti Gupta, Kimberly Newsome, Madelyn Baez-Santiago, Sally Slavinski, Jennifer L. White, Cynthia A. Moore, Carrie K. Shapiro-Mendoza, Lyle Petersen, Coleen Boyle, Denise J. Jamieson, Dana Meaney-Delman, Margaret A. Honein, and US Zika Pregnancy Registry Collaboration, "Vital Signs: Update on Zika Virus–Associated Birth Defects and Evaluation of All U.S. Infants with Congenital Zika Virus Exposure—U.S. Zika Pregnancy Registry, 2016," Morbidity and Mortality Weekly Report 66 (2017): 366–373.

3. EU Science Hub, "Climate Change Promotes the Spread of Mosquito- and Tick-Borne Viruses," news release, March 16, 2018.

4. USGCRP. The Impacts of Climate Change on Human Health in the United States: A Scientific Assessment, ed. A. Crimmins, J. Balbus, J. L. Gamble, C. B. Beard, J. E. Bell, D. Dodgen, R. J. Eisen, N. Fann, M. D. Hawkins, S. C. Herring, L. Jantarasami, D. M. Mills, S. Saha, M. C. Sarofim, J. Trtanj, and L. Ziska. (Washington: US Global Change Research Program, 2016).

5. Nate Seltenrich, "Safe from the Storm: Creating Climate-Resilient Health Care Facilities," Environmental Health Perspectives 126, no. 10 (2018): 102001.

6. Anemona Hartcollis and Nina Bernstein, "At Bellevue, a Desperate Fight to Ensure the Patients' Safety," New York Times, November 1, 2012.

7. CBS/AP, "NYC Hospital Successfully Evacuates 300 Patients after Superstorm Sandy," CBS News, November 1, 2012.

8. American College of Emergency Physicians, "Lessons Learned from Hurricane Sandy and Recommendations for Improved Healthcare and Public Health Response and Recovery for Future Catastrophic Events," December 22, 2015.

9. Brad Plummer, "Five Big Ways the United States Will Need to Adapt to Climate Change," *New York Times*, November 26, 2018.

10. Searchlight, "SVG Opens First Smart Hospital in Caribbean," *Interactive Media Ltd.*, September 10, 2013.

11. ABC 13, "How These Med Center Doors Could Prevent Millions of Dollars in Damage During a Flood," *ABC 13*, August 25, 2017.

12. Britni N. Riley, "Kayaking to Work: Two Nurses Hitched a Ride to Memorial Hermann—Texas Medical Center," *TMC Pulse* 4, no. 9 (2017): 13.

13. World Bank, "Climate-Smart Healthcare: Low-Carbon and Resilience Strategies for the Health Sector" (Washington, DC: World Bank, 2017), 34.

14. Khyati Kakkad, Michelle L. Barzaga, Sylvan Wallenstein, Gulrez Shah Azhar, and Perry E. Sheffield, "Neonates in Ahmedabad, India, During the 2010 Heat Wave: A Climate Change Adaptation Study," *Journal of Environmental and Public Health* (2014): 946875.

15. US Climate Resilience Toolkit, "Health Care Facilities Maintain Indoor Air Quality through Smoke and Wildfire," US Climate Resilience Toolkit, https://toolkit.climate.gov/case-studies/health-care-facilities-maintain-indoor-air-quality-through-smoke-and-wildfires.

16. Allison Linn, "Building a Better Mosquito Trap: How a Microsoft Research Project Could Help Track Zika's Spread," *Microsoft Features*, June 21, 2016.

17. Alan Boyle, "How Microsoft's Project Premonition Uses Robotic Traps to Zero in on Zika Mosquitoes," *GeekWire*, February 16, 2017.

18. Microsoft, "Technology Turns Mosquitoes into Allies Instead of Enemies in the Fight against Deadly Viruses," *Microsoft Transform*, January 26, 2017.

19. Barbara A. Han and John M. Drake, "Future Directions in Analytics for Infectious Disease Intelligence: Toward an Integrated Warning System for Emerging Pathogens," *EMBO Reports* 17, no. 6 (2016): 785–789.

20. Republic of South Africa Department of Science and Technology, "South Africa and Japan Boost Fight against Infectious Diseases," news release, August 7, 2014; Takayoshi Ikeda, Swadhin K. Behera, Yushi Morioka, Noboru Minakawa, Masahiro Hashizume, Ataru Tsuzuki, Rajendra Maharaj, and Philip Kruger, "Seasonally Lagged Effects of Climatic Factors on Malaria Incidence in South Africa," *Scientific Reports* 7 (2017): 2458.

21. Republic of South Africa Council on Scientific Research, "International Collaborative Effort Pinpoints Relationship between Climate and Malaria in Limpopo," news release, June 22, 2017.

22. Ikeda et al., "Seasonally Lagged Effects of Climatic Factors on Malaria Incidence in South Africa."

23. Amdavad Municipal Corporation, "Ahmedabad Heat Action Plan 2017: Guide to Extreme Heat Planning in Ahmedabad" (Ahmedabad, India: Amdavad Municipal Corporation, 2017), 2.
24. T. Shah, D. Mavalankar, P. Ganguly, P. Shekhar Dutta, A. Tiwari, A. Rajiva, A. Jaiswal, K. Knowlton, M. Connolly, N. Kaur, B. Deol, J. Hess, and P. Sheffield, "Innovative Heat Wave Early Warning System and Action Plan in Ahmedabad, India," in *Climate Services for Health: Improving Public Health Decision-Making in a New Climate*, ed. J. Shumake-Guillemot and L. Fernandez-Montoya (Geneva, Switzerland: World Health Organization and World Meteorological Organization, 2018): 162–165.
25. Shoumojit Banerjee, "IMD's New Weather Model Will Make Forecasts More Reliable," *The Hindu*, February 13, 2017.
26. Camilo Mora, Bénédicte Dousset, Iain R. Caldwell, Farrah E. Powell, Rollan C. Geronimo, Coral R. Bielecki, Chelsie W. W. Counsell, Bonnie S. Dietrich, Emily T. Johnston, Leo V. Louis, Matthew P. Lucas, Marie M. McKenzie, Alessandra G. Shea, Han Tseng, Thomas W. Giambelluca, Lisa R. Leon, Ed Hawkins, and Clay Trauernicht, "Global Risk of Deadly Heat," *Nature Climate Change* 7 (2017): 501–506.
27. Global Heat Health Information Network, "View Heat Health Action Plans," Global Heat Health Information Network, www.ghhin.org/map.
28. Brian S. Schwartz, Jason B. Harris, Ashraful I. Khan, Regina C. Larocque, David A. Sack, Mohammad A. Malek, Abu S. G. Faruque, Firdausi Qadri, Stephen B. Calderwood, Stephen P. Luby, and Edward T. Ryan, "Diarrheal Epidemics in Dhaka, Bangladesh, during Three Consecutive Floods: 1988, 1998, and 2004," *American Journal of Tropical Medicine and Hygiene* 74, no. 6 (2006): 1067–1073.
29. Kamran Reza Chowdhury, "Bangladesh Eliminates Post-Flood Disease Deaths," *The Third Pole*, August 30, 2017.
30. Amy Lieberman, "Q&A: Bangladesh Cholera Expert on How to Train Personnel Worldwide," *Devex*, June 26, 2018.
31. Rita Colwell, Anwar Huq, M. Sirajul Islam, K. M. A. Aziz, M. Yunus, N. Huda Khan, A. Mahmud, R. Bradley Sack, G. B. Nair, J. Chakraborty, David A. Sack, and E. Russek-Cohen, "Reduction of Cholera in Bangladesh Villages by Simple Filtration," *Proceedings of the National Academy of Sciences*, 100, no. 3 (2003): 1051–1055.
32. Lieberman, "Q&A: Bangladesh Cholera Expert on How to Train Personnel Worldwide."
33. Nick Watts, Markus Amann, Nigel Arnell, Sonja Ayeb-Karlsson, Kristine Belesova, Helen Berry, Timothy Bouley, Maxwell Boykoff, Peter Byass, Wenjia Cai, Diarmid Campbell-Lendrum, Jonathan Chambers, Meaghan Daly, Niheer Dasandi, Michael Davies, Anneliese Depoux, Paula Dominguez-Salas,

Paul Drummond, Kristie L. Ebi, Paul Ekins, Lucia Fernandez Montoya, Helen Fischer, Lucien Georgeson, Delia Grace, Hilary Graham, Ian Hamilton, Stella Hartinger, Jeremy Hess, Ilan Kelman, Gregor Kiesewetter, Tord Kjellstrom, Dominic Kniveton, Bruno Lemke, Lu Liang, Melissa Lott, Rachel Lowe, Maquins Odhiambo Sewe, Jaime Martinez-Urtaza, Mark Maslin, Lucy McAllister, Slava Jankin Mikhaylov, James Milner, Maziar Moradi-Lakeh, Karyn Morrissey, Kris Murray, Maria Nilsson, Tara Neville, Tadj Oreszczyn, Fereidoon Owfi, Olivia Pearman, David Pencheon, Steve Pye, Mahnaz Rabbaniha, Elizabeth Robinson, Joacim Rocklöv, Olivia Saxer, Stefanie Schütte, Jan C. Semenza, Joy Shumake-Guillemot, Rebecca Steinbach, Meisam Tabatabaei, Julia Tomei, Joaquin Trinanes, Nicola Wheeler, Paul Wilkinson, Peng Gong, Hugh Montgomery, and Anthony Costello. "The 2018 Report of the Lancet Countdown on Health and Climate Change: Shaping the Health of Nations for Centuries to Come," *The Lancet* 392, no. 10163 (2018): 2479–2514.

34. Tamma A. Carleton, "Crop-Damaging Temperatures Increase Suicide Rates in India," *Proceedings of the National Academy of Sciences of the United States of America* 114, no. 33 (2017): 8746–8751; Marshall Burke, Felipe González, Patrick Baylis, Sam Heft-Neal, Ceren Baysan, Sanjay Basu, and Solomon Hsiang, "Higher Temperatures Increase Suicide Rates in the United States and Mexico," *Nature Climate Change* 8 (2018): 723–729.

35. Ronald C. Kessler, Sandro Galea, Michael J. Gruber, Nancy A. Sampson, Robert J. Ursano, and Simon Wessely, "Trends in Mental Illness and Suicidality after Hurricane Katrina," *Molecular Psychiatry* 13 (2008): 374–384.

36. Nate Scott, "How One Couple's Desire to Rebuild New Orleans Nearly Tore Them Apart," *USA Today: For the Win*, August 28, 2015.

37. Susanta Kumar Padhy, Sidharth Sarkar, Mahima Panigrahi, and Surender Paul, "Mental Health Effects of Climate Change," *Indian Journal of Occupational and Environmental Medicine* 19, no. 1 (2015): 3–7.

38. Katie Hayes, G. Blashki, J. Wiseman, S. Burke, and L. Reifels, "Climate Change and Mental Health: Risks, Impacts and Priority Actions," *International Journal of Mental Health Systems* 12 (2018): 28.

39. Agence-France Presse, "Thousands of Farmer Suicides Prompt India to Set up $1.3bn Crop Insurance Scheme," *The Guardian*, January 14, 2016.

40. Susan Clayton, Christie Manning, Kirra Krygsman, and Meighen Speiser, "Mental Health and Our Changing Climate: Impacts, Implications, and Guidance" (Washington, DC: American Psychological Association and ecoAmerica, 2017), 49.

41. Renee Salas, Paige Knappenberger, and Jeremy Hess, "2018 Lancet Countdown on Health and Climate Change Brief for the United States of America," (London: The Lancet, 2018).

42. Margaret Chan, "WHO Director-General Addresses Event on Climate Change and Health" (speech, 2015 United Nations Climate Change Conference, Paris, France, December 8, 2015).

43. United Nations Environment Programme, "The Adaptation Gap Report 2018" (Nairobi, Kenya: United Nations Environment Programme, 2018), xiv.

Chapter 8

1. BBC, "Pope to Urge Swift Action on Global Warming," *BBC News*, June 16, 2015.

2. Pope Francis, *Laudato Si'* (Vatican City: Vatican Press, 2015), chap. 1, sec. 25.

3. Leonard A. Nurse, Graham Sem, J. E. Hay, A. G. Suarez, Poh Poh Wong, L. Briguglio, S. Ragoonaden, A. Githeko, J. Gregory, V. Ittekkot, U. Kaly, R. Klein, M. Lal, A. McKenzie, H. McLeod, N. Mimura, J. Price, Dahe Qin, B. Singh, and P. Weech, "Small Island States," in *Climate Change 2001: Impacts, Adaptation and Vulnerability. Contribution of Working Group II to the Third Assessment of the Intergovernmental Panel on Climate Change*, ed. J. J. McCarthy, Osvaldo F. Canziani, Neil A. Leary, David J. Dokken, and Kasey S. White (Cambridge, UK: Cambridge University Press, 2001).

4. International Monetary Fund, "The Effects of Weather Shocks on Economic Activity: How Can Low-Income Countries Cope?," in *Seeking Sustainable Growth: Short-Term Recovery, Long-Term Challenges* (Washington, DC: International Monetary Fund, 2017), 121.

5. Benjamin Sultan, "Global Warming Threatens Agricultural Productivity in Africa and South Asia," *Environmental Research Letters* 7, no. 4 (2016): 041001.

6. International Monetary Fund, "Effects of Weather," 121.

7. Marshall Burke, Solomon M. Hsiang, and Edward Miguel, "Global Non-linear Effect of Temperature on Economic Production," *Nature* 527 (2015): 235–239.

8. Solomon Hsiang, Robert Kopp, Amir Jina, James Rising, Michael Delgado, Shashank Mohan, D. J. Rasmussen, Robert Muir-Wood, Paul Wilson, Michael Oppenheimer, Kate Larsen, and Trevor Houser, "Estimating Economic Damage from Climate Change in the United States," *Science* 356, no. 6345 (2017): 1362–1369.

9. Jesse K. Anttila-Hughes and Solomon Hsiang, "Destruction, Disinvestment, and Death: Economic and Human Losses Following Environmental Disaster," Goldman School of Public Policy Working Paper, University of California, Berkeley, 2014.

10. Sharon Maccini and Dean Yang, "Under the Weather: Health, Schooling, and Economic Consequences of Early-Life Rainfall," *American Economic Review*, 99, no. 3 (2009): 1006–1026.

11. Board of Governors of the Federal Reserve System, "Report on the Economic Well-Being of U.S. Households in 2017" (Washington, DC: Board of Governors of the Federal Reserve System, 2018).

12. US Census Bureau, "2017 American Housing Survey" (Washington, DC: US Census Bureau, 2018).

13. Insurance Information Institute, "Number of Renters Is on the Rise—but Few of Them Have Insurance," news release, September 22, 2014.

14. Michel Masozera, Melissa Bailey, and Charles Kerchner, "Distribution of Impacts of Natural Disasters across Income Groups: A Case Study of New Orleans," *Ecological Economics* 63, no. 2–3 (2007): 299–306.

15. Mary Jo Gibson and Michele Hayunga, "We Can Do Better: Lessons Learned for Protecting Older Persons in Disasters" (Washington, DC: AARP Public Policy Institute, 2006).

16. Lex Frieden, "The Impact of Hurricanes Katrina and Rita on People with Disabilities: A Look Back and Remaining Challenges" (Washington, DC: National Council on Disability, 2006).

17. David M. Perry, "America Isn't Ready for Disability Disaster Response This Hurricane Season," *CityLab*, June 1, 2018.

18. Keenan, Hill, and Gumber, "Climate Gentrification."

19. Nicholas Kusnetz, "Norfolk Wants to Remake Itself as Sea Level Rises, but Who Will Be Left Behind?," *Inside Climate News*, May 21, 2018.

20. Zack Coleman and Daniel Cusick, "2 Towns, 2 Storms, and America's Imperiled Poor," *E&E News*, October 1, 2018.

21. Eric Klinenberg, *Heat Wave: A Social Autopsy of Disaster in Chicago* (Chicago: University of Chicago Press, 2002).

22. CIUSSS, "Canicule: Juillet 2018—Montréal Bilan Préliminaire" (Montréal, Canada: CIUSSS du Centre-Sud-de-l'Île-de-Montréal Direction Régionale de Santé Publique, 2018).

23. Eric Klinenberg, "Adaptation: How Can Cities Be 'Climate-Proofed'?," *New Yorker*, January 7, 2013.

24. Tatyana Deryugina, Laura Kawano, and Steven Levitt, "The Economic Impact of Hurricane Katrina on Its Victims: Evidence from Individual Tax Returns," NBER Working Paper No. 20713 (Cambridge, MA: National Bureau of Economic Research, 2014).

25. Vincent Mutie Nzau, "Insuring against Climate Risk in Kenya," *International Institute for Environment and Development Blog*, June 6, 2017.

26. City of New York, "A Stronger, More Resilient New York" (New York, NY: City of New York, 2013).

27. City of New York, "One New York: The Plan for a Strong and Just City" (New York: City of New York, 2015).

28. Stephane Hallegatte, Adrien Vogt-Schilb, Mook Bangalore, and Julie Rozenberg, "Unbreakable: Building the Resilience of the Poor in the Face of Natural Disasters" (Washington, DC: World Bank, 2016).

29. William Alden, "Around Goldman's Headquarters, an Oasis of Electricity," *New York Times*, November 2, 2012.

30. Sheeraz Raza, "Goldman Sachs' Gary Cohn Talks about the Impact of Sandy on the Bank," *ValueWalk*, October 30, 2012.

Chapter 9

1. Hannah Strange, "Super Typhoon Haiyan Smashes into Philippines," *The Telegraph*, November 8, 2013.

2. Isabel Makhoul, "Recovery and Return after Typhoon Haiyan/Yolanda," in *The State of Environmental Migration 2014: A Review of 2013*, ed. François Gemenne, Pauline Brücker, and Dina Ionesco (Le Grand-Saconnex and Paris, France: International Organization for Migration and Sciences Po, 2014), 19.

3. Damien McElroy, "Philippine President Makes Frank Admission of Typhoon Haiyan Relief Failures," *The Telegraph*, November 18, 2013.

4. Alice R. Thomas, "Resettlement in the Wake of Typhoon Haiyan in the Philippines: A Strategy to Mitigate Risk or a Risky Strategy?" (Washington, DC: Brookings Institution, 2015).

5. Ibid.

6. David Doyle, "Rebuilding after Typhoon Haiyan: 'Every Time There Is a Storm I Get Scared,'" *The Guardian*, October 3, 2016.

7. Office of the Director of National Intelligence, "Implications for US National Security of Anticipated Climate Change" (Washington, DC: National Intelligence Council, 2016), 7.

8. Kumari Rigaud, Alex de Sherbinin, Bryan Jones, Jonas Bergmann, Viviane Clement, Kayly Ober, Jacob Schewe, Susana Adamo, Brent McCusker, Silke Heuser, and Amelia Midgley, "Groundswell: Preparing for Internal Climate Migration" (Washington, DC: World Bank, 2018).

9. Elizabeth Fleming et al., "Coastal Effects," in *Impacts, Risks, and Adaptation in the United States: Fourth National Climate Assessment*, vol. 2, ed. D. R. Reidmiller, C. W. Avery, D. R. Easterling, K. E. Kunkel, K. L. M. Lewis, T. K. Maycock, and B. C. Stewart (Washington, DC: US Global Change Research Program, 2018), 335.

10. National Research Council, *Climate and Social Stress: Implications for Security Analysis* (Washington, DC: The National Academies Press, 2013), 112–117.

11. Rex Victor Cruz, Hideo Harasawa, Murari Lal, Shaohong Wu, Yurij Anokhin, Batima Punsalmaa, Yasushi Honda, Mostafa Jafari, Congxian Li, and Nguyen Huu Ninh, "Asia," in *Climate Change 2007: Impacts, Adaptation and*

Vulnerability. Contribution of Working Group II to the Fourth Assessment Report of the Intergovernmental Panel on Climate Change, ed. Martin, Parry Osvaldo Canziani, Jean Palutik, Paul van der Linden, and Clair Hanson (Cambridge, UK: Cambridge University Press, 2007), 492.

12. Ezekiel Simperingham, "Climate Displacement in Bangladesh The Need for Urgent Housing, Land and Property (HLP) Rights Solutions" (Geneva, Switzerland: Displacement Solutions, 2012).

13. Human Rights Watch, "'Trigger Happy': Excessive Use of Force by Indian Troops at the Bangladesh Border" (New York, NY: Human Rights Watch, 2010), 8.

14. Mathew E. Hauer, "Migration Induced by Sea-Level Rise Could Reshape the US Population Landscape," *Nature Climate Change* 7 (2017): 321, 324.

15. National Oceanic and Atmospheric Administration, "What Is Sea Level? And Why Is the Sea Not Exactly Level?," *NOAA News & Features*, August 10, 2017.

16. Kristina A. Dahl, Erika Spanger-Siegfried, Astrid Caldas, and Shana Udvardy, "Effective Inundation of Continental United States Communities with 21st Century Sea Level Rise," *Elementa: Science of the Anthropocene* 5 (2017): 37.

17. Fleming, "Coastal Effects," 334.

18. Iowa Homeland Security and Emergency Management, "2008 Iowa Mitigation Success Story – Avoided Losses through Property Acquisition and Relocation for Open Space" (Windsor Heights, IA: Iowa Homeland Security and Emergency Management, 2008).

19. Richard Turner Henderson, "Sink or Sell: Using Real Estate Purchase Options to Facilitate Coastal Retreat," *Vanderbilt Law Review* 71 (2018): 64180.

20. Lisa Song, Al Shaw, and Neena Satija, "After Harvey, Buyouts Won't Be the Answer for Frequent Flood Victims in Texas," *Texas Tribune*, November 2, 2017.

21. Regeneris Consulting, "Coastal Pathfinder Evaluation: An Assessment of the Five Largest Pathfinder Projects" (London: UK Department for Environment Food and Rural Affairs, 2011).

22. Darryl Fears, "Built on Sinking Ground, Norfolk Tries to Hold Back Tide amid Sea-Level Rise," *Washington Post*, June 17, 2012.

23. Ryan Murphy, "Norfolk Is Fighting Flooding by Giving Part of the City Back to Nature," *Virginian-Pilot*, October 29, 2018.

24. California Coastal Commission, "Residential Adaptation Policy Guidance: Interpretive Guidelines for Addressing Sea Level Rise in Local Coastal Programs" (San Francisco: California Coastal Commission, 2018), 28.

25. Bianca Kaplanek, "'Managed Retreat' Nixed in Sea-Level Rise Plan," *Coast News Group*, May 25, 2018.

26. Anne C. Mulkern, "City Vows to Fight State over 'Retreat,'" *E&E News*, October 17, 2018.

27. Matthew Glass, *Ultimatum* (New York: Atlantic Monthly Press, 2009).

28. Asian Development Bank, "A Region at Risk: The Human Dimensions of Climate Change in Asia and the Pacific" (Manila: Asian Development Bank, 2017), 82.

29. National Research Council, *Climate and Social Stress*, 112–117.

30. Peter Grier, "The Great Katrina Migration," *Christian Science Monitor*, September 12, 2005.

31. Laura Bliss, "10 Years Later, There's So Much We Don't Know about Where Katrina Survivors Ended Up," *CityLab*, August 15, 2015.

32. Molly Fifer McIntosh, "Measuring the Labor Market Impacts of Hurricane Katrina Migration: Evidence from Houston, TX" (paper presented at the Annual Meeting of the Allied Social Science Associations, New Orleans, LA, January 6, 2008).

33. Sara Chaganti and Jasmine Waddell, "Employment Change among Hurricane Katrina Evacuees: Impacts of Race and Place," *Journal of Public Management & Social Policy* 22, no. 2 (2015): 3.

34. Angela Sherwood, Megan Bradley, Lorenza Rossi, Rufa Guiam, and Bradley Mellicker, "Resolving Post-Disaster Displacement: Insights from the Philippines after Typhoon Haiyan (Yolanda)" (Washington, DC: Brookings Institution and International Organization for Migration, 2015), 30–31.

35. Mathew E. Hauer, "Migration Induced by Sea-Level Rise Could Reshape the US Population Landscape," *Nature Climate Change* 7 (2017): 321–325.

36. GIZ, "Support for Climate Migrants," Deutsche Gesellschaft für Internationale Zusammenarbeit, https://www.giz.de/en/mediacenter/58801.html.

37. Saleemul Huq, "Building Climate Resilient, Migrant-Friendly Cities," *Daily Star*, March 28, 2018.

38. Laurie Goering, "Rising Tide of Climate Migrants Spurs Dhaka to Seek Solutions," Reuters, April 26, 2016.

39. Steve Baragona, "Miami Faces Future of Rising Seas," *Voice of America*, November 22, 2017.

40. Christine Lagarde, "Migration: A Global Issue in Need of a Global Solution," *IMFBlog*, November 11, 2015.

41. Bliss, "10 Years Later.".

42. Robin Bronen, "Community-Based Adaptation: Alaska Native Communities Design a Relocation Process to Protect Their Human Rights," in *Resilience: The Science of Adaptation to Climate Change*, ed. Zinta Zommers and Keith Alverson (Amsterdam, Netherlands: Elsevier, 2018), 117.

43. Anouch Missirian and Wolfram Schlenker, "Asylum Applications Respond to Temperature Fluctuations," *Science* 358, no. 6370 (2017): 1610–1614.

44. Jonathan Pearlman, "New Zealand Creates Special Refugee Visa for Pacific Islanders Affected by Climate Change," *Straits Times*, December 9, 2017.

Chapter 10

1. Catherine Schkoda, Shawna Cuan, and E. D. McGrady, "Proceedings and Observations from a Climate Risk Event March 19–20, 2015" (Arlington, VA: CNA, 2015), 2.
2. Ibid.
3. Office of the Director of National Intelligence, "Implications for US National Security of Anticipated Climate Change," 3.
4. American Security Project, "The Global Security Defense Index on Climate Change" (Washington, DC: American Security Project, 2014).
5. T. A. Middendorp, "Toespraak van generaal T.A. Middendorp, bij dhet Planetary Security Initiative" (speech, Planetary Security Initiative, The Hague, December 5, 2016).
6. David Eckstein, Vera Künzel and Laura Schäfer, "Global Climate Risk Index 2018" (Bonn, Germany: Germanwatch e.V., 2017).
7. Nathaniel Gronwold, "What Caused the Massive Flooding in Pakistan?," *Scientific American*, October 12, 2010.
8. Sualiha Nazar, "Pakistan's Big Threat Isn't Terrorism—It's Climate Change," *Foreign Policy*, March 4, 2016.
9. Syed Muhammad Abubakar, "Pakistan 7th Most Vulnerable Country to Climate Change, Says Germanwatch," *Dawn*, November 9, 2017.
10. Anwar Iqbal, "Water Talks Failed to Produce Agreement in Pakistan's Water Dispute with India: WB," *Dawn*, May 24, 2018.
11. Asma Khan Lone, "How Can Climate Change Trigger Conflict in South Asia?," *Foreign Policy*, November 20, 2015; Ruth Schuster, "Pakistan's Deadly Paradox: It Can't Combat Climate Change, but It Has No Choice," *Haaretz*, July 16, 2018.
12. Gronwold, "What Caused the Massive Flooding in Pakistan?"
13. Saeed Shah, "Pakistan Flood Response Prompts Rising Anti-Government Resentment," *The Guardian*, August 13, 2010; Thomas D. Kirsch, Christina Wadhwani, Lauren Sauer, Shannon Doocy, and Christina Catlett, "Impact of the 2010 Pakistan Floods on Rural and Urban Populations at Six Months," *PLoS Currents* 4 (2012): e4fdfb212d2432.
14. The Editors of Encyclopaedia Britannica, "Pakistan Floods of 2010," *Encyclopaedia Britannica*, March 13, 2012.
15. Hasnain Kazim, "Taliban Courts Pakistan Flood Victims, Race to Provide Aid Emerges between West and Extremists," *Der Spiegel*, August 16, 2010.
16. John Moreau, "How the Pakistani Floods Help the Taliban," *Newsweek*, August 12, 2010.
17. Rob Crilly, "Pakistan Flood Aid from Islamic Extremists," *The Telegraph*, August 21, 2010.

18. Dave Philipps, "Exposed by Michael: Climate Threat to Warplanes at Coastal Bases," *New York Times*, October 17, 2018.

19. Ibid.

20. Bilal M. Ayyub, Haralamb G. Braileanu, and Naeem Qureshi, "Prediction and Impact of Sea Level Rise on Properties and Infrastructure of Washington, DC," *Risk Analysis* 32, no. 11 (2012): 1901–1918.

21. Justin Nobel, "What Happens When a Superstorm Hits D.C.?" *Rolling Stone*, September 21, 2017.

22. Matt Connolly, "5 Things to Know about Hurricanes, Hampton Roads and National Security," *Center for Climate and Security Blog*, October 2, 2015, https://climateandsecurity.org/2015/10/02/5-things-to-know-about-hurricanes-hampton-roads-and-national-security/.

23. Camilo Mora, Daniele Spirandelli, Erik C. Franklin, John Lynham, Michael B. Kantar, Wendy Miles, Charlotte Z. Smith, Kelle Freel, Jade Moy, Leo V. Louis, Evan W. Barba, Keith Bettinger, Abby G. Frazier, John F. Colburn IX, Naota Hanasaki, Ed Hawkins, Yukiko Hirabayashi, Wolfgang Knorr, Christopher M. Little, Kerry Emanuel, Justin Sheffield, Jonathan A. Patz, and Cynthia L. Hunter, "Broad Threat to Humanity from Cumulative Climate Hazards Intensified by Greenhouse Gas Emissions," *Nature Climate Change* 8 (2018): 1062–1071.

24. Office of the Director of National Intelligence, "Implications for US National Security of Anticipated Climate Change," 5.

25. Hu Xi, "China Dialogue Special Report: Climate Change Poses Grave Threats to China's Essential Infrastructure 2016," *China Dialogue*, April 20, 2016.

26. Dave Hill, "Beyond the Thames Barrier: How Safe Is London from Another Major Flood?," *The Guardian*, February 19, 2015.

27. John Vandiver, "50,000 Troops to Drill for Invasion in Largest Europe Exercise since Cold War," *Stars and Stripes*, October 24, 2018.

28. Malte Humpert, "The Future of Arctic Shipping: A New Silk Road for China?" (Washington DC: Arctic Institute Center for Circumpolar Security Studies, 2013), 5.

29. Robert Woodward, *Obama's Wars* (New York: Simon and Schuster, 2011), 210.

30. Ibid.

31. Barack Obama, "Memorandum on Climate Change and National Security" (Washington, DC: US Government Publishing Office, 2016).

Conclusion

1. Wake Smith and Gernot Wagner, "Stratospheric Aerosol Injection Tactics and Costs in the First 15 Years of Deployment," *Environmental Research Letters* 13, no. 12 (2018): 124001.

2. Anthony C. Jones, James M. Haywood, Nick Dunstone, Kerry Emanuel, Matthew K. Hawcroft, Kevin I. Hodges, and Andy Jones, "Impacts of Hemispheric Solar Geoengineering on Tropical Cyclone Frequency," *Nature Communications* 8 (2017): 13828.

3. Jonathan Proctor, Solomon Hsiang, Jennifer Burney, Marshall Burke, and Wolfram Schlenker, "Estimating Global Agricultural Effects of Geoengineering Using Volcanic Eruptions," *Nature* 560, no. 7719 (2018): 480.

4. Anthony Leiserowitz et al., "Climate Change in the American Mind."

5. Ibid.

6. Council on Climate Preparedness and Resilience, "Opportunities to Enhance the Nation's Resilience to Climate Change."

INDEX

For the benefit of digital users, indexed terms that span two pages (e.g., 52–53) may, on occasion, appear on only one of those pages.